Upskilling the Chinese workforce?

Hao Zhang

Upskilling the Chinese workforce?

An Institutional Dilemma in China's Vocational Education System

palgrave
macmillan

Hao Zhang
Renmin University of China
Beijing, China

ISBN 978-981-95-2655-0 ISBN 978-981-95-2656-7 (eBook)
https://doi.org/10.1007/978-981-95-2656-7

This Palgrave Macmillan imprint is published by the registered company Springer Nature Singapore Pte Ltd.
The registered company address is: 152 Beach Road, #21-01/04 Gateway East, Singapore 189721, Singapore

To my parents,
my doctoral advisors,
and all who shared their time, insights, and support for this work.

Competing Interests The author has no competing interests to declare that are relevant to the content of this manuscript.

Contents

List of Figures

List of Tables

CHAPTER 1

National Skill Formation Systems: Theory and International Experience

China's "economic miracle" over the past four decades has been largely based on the low-skilled, labor-intensive manufacturing sector boosting export revenues. This developmental strategy has been challenged in recent years, however, given slowing economic growth, rising labor costs and conflicts, and a shortage of skilled workers. The World Bank (2013) has encouraged the Chinese government to restructure and upgrade its industries to avoid a middle-income trap. Global post-war developmental experiences have indicated that vocational workforce skills are crucial for sustainable economic growth and international competitiveness in both developed (Streeck, 1992) and emerging market economies (Ashton et al., 1999). Upskilling the workforce has thus become a crucial aspect of China's recent economic restructuring efforts.

The vocational education and training (VET) system has thereby become a recent focal point of China's economic reform, gaining growing importance in advancing the skill development of the Chinese workforce. Figure 1.1 illustrates the trend in the total number of vocational school graduates and the estimated proportion of those employed upon graduation within the total urban workforce of China since the turn of the century. Clearly, vocational schools have become increasingly important labor providers in the economy. In light of this, not only have employers been engaging vocational schools to tackle skilled labor shortages (Li & Sheldon, 2010), but the state has also implemented a series of reforms in

H. Zhang, *Upskilling the Chinese workforce?*,
https://doi.org/10.1007/978-981-95-2656-7_1

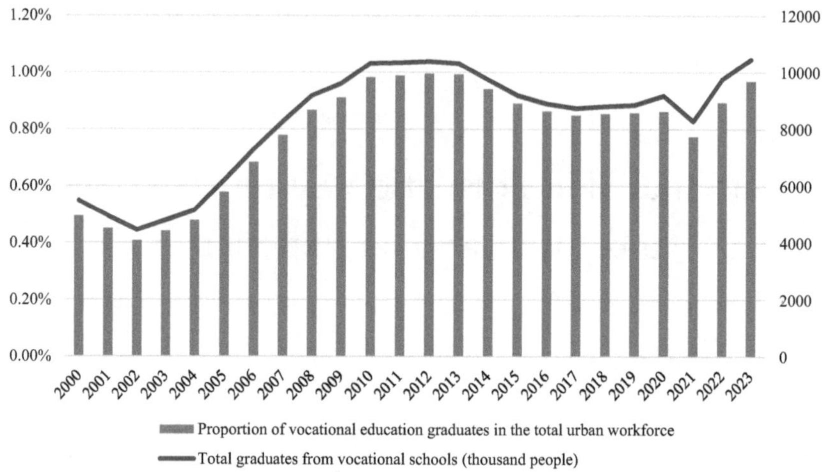

Note: estimates based on official data. See Appendix A for detailed data and the estimation method.

Fig. 1.1 Total vocational school graduates and their proportion in the total urban workforce since the turn of the century

recent years to facilitate a profound industrial restructuring and upgrading in China, which includes an ambitious workforce upskilling agenda.

Employers are now engaging in extensive collaboration with vocational schools to address the problem of skilled labor shortage. Literature (Li & Sheldon, 2014) and my own research have revealed school-firm collaborative strategies including a) providing various forms of financial support, (2) transferring the latest technology and know-how to vocational schools, (3) co-developing curricula and training materials, (4) the co-building on-campus practical training centers, (5) dispatching engineers and technicians to teach at schools, and (6) facilitating training for schoolteachers in their own workplaces and corporate training facilities.

Additionally, a profound reform of the Chinese VET system has been placed on the state's agenda. The "Made in China 2025" initiative, issued by the central government in 2015, announced continued investment in human capital within the Chinese workforce, as well as skill development reform agendas, as vital institutional supports for industrial restructuring and upgrading. The overall objective was to sustain the Chinese "economic miracle" into the forthcoming decades. Leveraging the VET system was deemed by the guideline a vitally important component of the central government's "supply-side reform," which sought to leverage workers' skill development processes in order to encourage skill and industrial upgrading by employers. As of now—ten years after the implementation of the initiative—China's fully fledged industrial sector clearly constitutes a key competitive advantage and strategic asset in the post-pandemic era of international competition.[1] It is therefore an opportune moment to look back and reflect on the reform initiated a decade ago and to assess what has been done well and what remains problematic. This book aims at this research-based policy evaluation agenda, focusing specifically on the Chinese VET system.

Although media reports have by and large covered these recent developments in the Chinese VET system, scholarly understanding of this topic remains inadequate and sometimes biased as of now. Our inquiry here is situated at the confluence of two strands of literature.[2] In the institutional literature, prevailing wisdom depicts the Chinese VET system as primarily school-based and state-centralized (Cooke, 2005), requiring central and local governments to tightly control the VET process. The system leaves schools very little room to devise their own programs and does not offer any formal and stable institutional mechanisms to incorporate firms' input into the training process. This has rendered the VET process in China ineffective and non-market oriented, and consequently, it often fails to accommodate the skill needs of employers (Durden & Yang, 2006). Although a decentralization reform has been implemented since the mid-1980s with the aim of promoting school-firm collaboration, drawing on empirical evidence from different fragmentations of the Chinese VE system, scholars have found a lack of strong, institutionalized school-firm ties (Lai & Lo, 2006), as well as a failure to satisfy students' needs in those

[1] See, for instance, Kyle Chan's guest essay on The New York Times on May 29, 2015, available at: https://www.nytimes.com/2025/05/19/opinion/china-us-trade-tariffs.html

[2] There is another important strand of scholarship—primarily emerging from the field of education—that falls beyond the scope of this book. This body of work generally adopts a micro-level, pedagogical perspective, focusing either on descriptive analyses of the current landscape of China's vocational education and training (VET) system (Wu & Ye, 2018; Xue & Li, 2022) or, using micro-level data, on empirical assessments of educational practices and student outcomes in vocational education (Xu, 2023; Yi et al., 2013, 2018). While these studies offer valuable insights into classroom dynamics and educational effectiveness, their analytical focus tends to center on implementation and pedagogy rather than on broader institutional challenges of the VET reforms. As a result, although complementary in scope, this literature addresses different sets of questions than those guiding the central inquiry of this book.

programs (Muller, 2024; Yi et al., 2018). This, associated with the expansion of China's entire education system during the same period, has led to skill discrepancies between vocational schools and industry (Song & Xu, 2024).

In another strand of literature, ethnographic research revealed ineffective training processes within vocational schools' classrooms, such that students barely acquire useful skills, but rather construct an identity of "valued commodity" (Woronov, 2012). Critical sociologists found that student interns in an electronics manufacturing firm were treated as unskilled, seasonal, cheap labor. They thus concluded that Chinese vocational schools are quasi-labor agencies that simply trade student interns with firms in exchange for unjustified commissions (Chan et al., 2015; Smith & Chan, 2015; Song, 2024; Su, 2010–2011). This depiction of labor agencies has become the dominant perception of Chinese vocational schools among many scholars, even though this research is contextualized in a potentially biased setting. The "labor agency" narrative of Chinese vocational schools emerged from a research project that targeted Foxconn. In addition to the obvious industry bias, it is unclear whether the authors implemented a sound school sampling strategy. I argue that these stereotypes correctly reflect several existing problems within a certain portion of Chinese vocational schools, but they have not revealed the full scenario of the Chinese VET system.

Both strands of literature on Chinese VET have left unresolved the puzzle of secondary labor market failure. The literature implies that school-firm collaboration in China is a secondary labor market strategy for employers: unable to produce useful skilled workers, schools function as temporary agencies that supply flexible, cheap labor to collaborative firms. In this book, however, I found that employers engage in a wide portfolio of collaborative activities with vocational schools, including various forms of technology and knowledge transfer to schools—often at no cost to the schools. If all these practices are merely secondary labor market strategies aimed at securing unskilled seasonal labor, then why would firms not simply rely on existing labor agencies (which have been sufficiently developed currently in China and could provide large scales of cheap labor to employers) that do not entail these additional costs?

More importantly, as is widely known, firms in China over the past couple of decades have faced a major labor shortage. This is a structural shortage of skilled workers: there remains a surplus of relatively low-skilled migrant workforce (Kwan, 2009), but employers often cannot secure technicians and other skilled workers in the labor market (Chan, 2010; Farrell & Grant, 2005; Li & Sheldon, 2010). This structural shortage implies a major secondary labor market failure—that is, labor market institutions (e.g., temp agencies) cannot effectively solve the most urgent skilled labor shortage problem. If the hypothesis advanced in the literature—that vocational schools primarily function as labor agencies rather than effective skill formation institutions—is correct, then what has motivated employers to invest significant time and resources in the VET process? In other words, it is reasonable to suspect that vocational schools provide something more than what labor agencies provide—something that constitutes at least a temporary, partial fix to the secondary labor market failure.

This book seeks to resolve this puzzle of secondary labor market failure by carefully examining the Chinese VET reforms since the economic reform and opening-up, and the interaction modes among key players in VET, including schools, employers, and the state. Broadly speaking, I am interested in what is unique about the VET system in China and how and why the system operates and has been reformulated. What forces actively shape this process? In particular, now that both employers and the state have been incentivized to leverage the VET system for various political and economic reasons, which of them dominates the actual training process? Similarly, in the arena of school development and reform, what does the division of labor and cost-sharing mechanism between the state and employers look like? In practical terms, I aim to evaluate the capacity of the current Chinese VET system to upskill the Chinese workforce in the forthcoming decades and to identify and examine the challenges currently facing the skill-upgrading agenda.

As noted, the Chinese central government has been taking measures to decentralize a school-based national skill development system into a more marketized model that integrates extensive employer input since the mid-1980s. Under this decentralization reform, the state has granted substantial authority to schools to collaborate with firms and devise customized training programs for them. During the past decade, this decentralization reform has been further deepened by the two ministries with jurisdiction over the Chinese VET system—the Ministry of Education (MOE) and the Ministry of Human Resources and Social Security (MOHRSS)—through their separate apprenticeship reforms, modeled after the German vocational training system. Some of those collaborative training programs have eventually been institutionalized within the apprenticeship reforms. I find, however, that this distinctive targeted training model has been guided primarily by the short-term, specific skill needs of individual firms and that there is a lack of independent, industry-level skill standards for cross-program coordination. I contend that this disorganized skill development system is a product of the absence of effective private governance. What is needed, in other words, is the participation of employer associations and labor unions to coordinate the diversified interests of firms at the industry level in order to facilitate a successful workforce upskilling agenda in the long run.

In the following section of this chapter, I delineate two theoretical problems concerning skill formation. The first arises from the strategic human resource management literature on skills acquisition. This is relevant to the Chinese VET system as it raises a theoretical paradox: on the one hand, the stereotypical depiction of Chinese vocational schools as labor agencies, and on the other, the failure of the secondary labor market to provide useful skilled labor to firms. To resolve this paradox, I draw on a second stream of literature—namely, human capital theory and the institutional analysis of national skill formation systems. Based on this foundation, I have developed a theoretical model that affords a more sophisticated assessment of the recent Chinese apprenticeship reforms in VET.

Emergence and Failure of the Secondary Labor Market in China

Human capital theory distinguishes between general and firm-specific skills. General skills are considerably transferable across firms or industries, while firm-specific skills are useful only within a particular employer and, as a result, are often scarce in the labor market and acquirable only through internal training (Davis-Blake & Uzzi, 1993). Firms therefore must invest in developing firm-specific skills but are often discouraged from investing in general skills due to the potential for free-riding in the labor market, where other firms might poach trained workers.

Firms traditionally relied on internal labor markets to generate the specific skills they needed (Piore, 2002). An internal labor market is defined as "an administrative unit, such as a manufacturing plant, within which the pricing and allocation of labor is governed by a set of administrative rules or procedures" (Doeringer & Piore, 1971). It typically denotes a modern employment organization in which workers climb career ladders, and their wages and promotions are determined by organizational customs, norms, and power structures rather than by supply-demand equilibria. Internal labor market theory provides a socio-political explanation for employment outcomes within organizations that goes beyond purely economic doctrines. Workplace rules and practices—particularly managerial strategies—shield employees from external market competition by providing them with a certain level of job security and wage premiums (Osterman & Burton, 2006).

This internal labor market model, based on the historical experience of Western countries, provides for the long-term, internal cultivation of firm-specific skills, relying essentially on three important mechanisms. First, internal skills formation heavily relies on extensive informal, on-the-job training to secure crucial firm-specific skills (Klein et al., 1978; Williamson, 1981). These skills are usually unavailable from external, institute-based training programs, as they are not required by a sufficiently wide range of employers (Williamson et al., 1975). Second, firms sometimes need to incorporate seniority into their pay systems to make on-the-job training effective—only in this way will skilled workers be willing to train youth and sometimes apprentices without undue concern about being undercut by newcomers (Doeringer & Piore, 1971). Finally, firms also need to offer job security and internal promotion opportunities to retain these trained, skilled workers, often under negotiated agreements with labor unions (Knoke & Ishio, 1994). This arrangement wins young workers' consent to defer the return on their human capital to later stages of their careers, in exchange for the learning opportunities provided by seniors.

Yet the 1980s and 1990s witnessed a major shift in skill strategies across many industries worldwide, moving away from internal labor market models toward external skill acquisition. To some extent, this shift was driven by increasing competitive pressures due to globalization and the growing need for employment flexibility (Cappelli, 1995). Firms thus faced a dilemma: whether to develop skills and provide the training internally ("make") or acquire them from outside ("buy") through various labor market intermediaries, such as temp agencies, business partners, VET institutions, and others. A partial solution was to adopt a more differentiated approach—applying either "make" or "buy" strategies to different segments of the workforce, depending on their value and skill specificity (Osterman, 1982; Lepak & Snell, 1999). Firms have faced a "make-or-buy" dilemma in skills acquisition. However, this strategy—centered on retaining only a small core of internally trained workers—does not appear to meet the needs of many manufacturing and service-sector firms (Luo et al., 2021). For these firms, frontline operational workers are essential, and relying heavily on a secondary labor market model is insufficient for developing the firm-specific skills required.

This shift from internal labor markets to external skill acquisition, observed in advanced economies during the 1980s and 1990s, also took place in China—mainly through the rise of secondary labor markets that supplied cheap labor, consistent with China's early industrialization goals. Post-socialist labor reforms featured the "casualization of employment" and the breaking of the "iron rice bowl" (Friedman & Lee, 2010). This went hand in hand with a fundamental shift in employers' HR strategies—from internal to secondary labor markets—evident in the shrinking internal labor markets in the state sector and the emergence and expansion of an informal sector (Gallagher et al., 2011). Rising legal costs associated with new pro-labor legislation since 2008 further encouraged employers to externalize employment, turning to secondary labor market mechanisms such as labor dispatch agencies (Feng, 2019; Liu, 2015; Xu, 2008). Economists estimate that employment in China's informal sector accounts for over 64% of the total workforce (Zhou, 2013).

Vocational Schools as Labor Agencies?

Existing scholarly understanding of the Chinese VET system is inadequate and somewhat biased. Early studies focused on national-level analysis, portraying Chinese vocational schools as ineffective and non-market-oriented, attributing these shortcomings to state centralization and a rigid VET system (Cooke, 2005; Durden & Yang, 2006; Lai & Lo, 2006). As a result, vocational schools were seen as largely failing to meet industry skill demands. Later, critical researchers conducted fieldwork at Foxconn—a major electronics manufacturer—and found widespread use of student interns as temporary workers during peak seasons. Under the label of "interns," these students were often assigned to routine assembly-line tasks, supervised jointly by school teachers and factory managers (Chan et al., 2015; Smith & Chan, 2015; Song, 2024). Su (2010–2011) concluded that school-firm collaboration in China reflects a dual commodification of labor and education: through internships, schools supply firms with flexible, low-cost labor—a practice that has become an important, yet ethically questionable, source of revenue. Given their failure to produce skilled workers on one hand, and their role in supplying cheap seasonal labor on the other, Chinese vocational schools have long been stereotyped as functioning no differently from labor agencies. While Su called them "middlemen," media reports have labeled them "illegal labor brokers" (*hei zhongjie* 黑中介).[3]

This literature, however, has methodological limitations. Early research on the VET system relied on limited and often secondary data, much of it collected before or around the end of the twentieth century. Given the deepening marketization of the VET sector over the past decade, the current reality may differ significantly. Studies focusing on student internships are also narrow in scope—largely confined to the electronics manufacturing industry, which, as Lüthje et al. (2013) note, relies heavily on standardized mass production and thus has more assembly-line jobs than many other sectors. Indeed, much of this research stems from a single collaborative project centered on Foxconn. With only one employer as a focal point, it remains unclear whether the selection of vocational schools followed a representative sampling strategy. This narrow and potentially biased portrayal leaves room for a more nuanced analysis. It is reasonable to expect considerable variation among Chinese vocational schools—variation that goes beyond the "labor agency" stereotype.

Admittedly, some Chinese vocational schools may have historically functioned as secondary labor market institutions, supplying low-skill, low-cost labor. In Chap. 2, however, I find far greater diversity in the Chinese VET system than previously recognized. Based on firsthand research across a diverse sample of vocational schools and their partner firms in Guangdong, I identify four models of skill development: the ***high-performance model***, the ***industry-focused model***, the ***local market-oriented model***, and the ***labor agency model*** and provide explanations.

Furthermore, as rising labor costs have pushed Chinese firms toward more value-added production, it has become increasingly clear that vocational schools often fail to provide adequately trained workers (Cooke, 2005; Durden & Yang, 2006). Inspired by Western models—especially the German dual apprenticeship system—the central goal of VET reform in China has been to bring industry actors into the training process, promoting school-firm collaboration to better meet industry needs (Lai & Lo, 2006). To date, I find the state has largely withdrawn from the day-to-day operations of vocational schools, granting them autonomy to design and implement training programs, while encouraging employers to

[3] http://news.youth.cn/jy/201604/t20160419_7879083.htm, March 16, 2017.

contribute to curriculum development. This decentralization has laid the institutional foundation for the targeted training model—a distinctive feature of China's current VET system, which I explore in Chap. 3.

Specifically, firms are moving beyond the secondary labor market strategy and seeking to establish more meaningful collaborations with vocational schools to acquire the highly skilled workers that they need. Scholars have documented such collaborative training practices using labor market survey data (Li & Sheldon, 2014). Recent studies have paid more attention to the micro-level processes of these practices and found various problems including the low quality of the training processes (Woronov, 2016), the lack of consideration of students' needs (Müller, 2024), and the continued discrepancies between VET and the skill demands of the industry (Song & Xu, 2024). However, these scholars have generalized their arguments from a limited number of case studies, and thus by and large overlooked the variations within these partnerships and the lessons we can draw from relatively successful cases. It is theoretically difficult for firms to pursue numerical flexibility and labor equipped with sufficient firm-specific skills at the same time without innovative formats of school-firm collaboration and cost-sharing among a wide portfolio of stakeholders. While such institutional frameworks have been historically institutionalized in exemplar Western and Asian skill development models, we know very little about how collaborative training works in China and the burden of human capital investment is actually shared at the micro level. In Chap. 3, I examine the genesis, development, and performance of school-firm partnerships in the auto maintenance, robotics manufacturing, and beauty service industries in Guangdong province, with a view to learning about what causes success or failure and to examining the potential to transfer these practices to other parts of China and beyond.

This bottom-up experimentation with collaborative, targeted training has been institutionalized by the central state in recent years within two major apprenticeship reforms. Chapter 4 is tasked with the mission of providing a critical policy evaluation of these reforms, drawing on human capital theory and the literature on the post-war Western and Asian experience in skill development. In the following part of this chapter, I conduct a theoretical preparation for this research and juxtapose national skills formation system models around the world. International experience has suggested that several institutional arrangements are needed for skill formation models to be successful. Based on this discussion, I establish a theoretical framework that is useful for assessing the abovementioned reform outcomes in China.

National Skill Formation Systems

To begin with, the central mission of a skill formation system is to promote an appropriate cost-sharing structure to effectively engage employers and individual trainees (or households) for the sake of sufficient investment in human capital (Hashimoto, 1981). Typically, the state stipulates coherent institutional arrangements to frame this structure, while employers and individuals are motivated by various institutional and economic incentives to co-share the costs in different formats. Getting employers to pay is both a key and a difficult task. Policymakers and practitioners of advanced Western and Asian skill development systems—in particular, the state and labor unions—historically have made tremendous political and economic efforts to engage employers in skill development (Thelen, 2004). In this section, I summarize the key features of two ideal national skill formation systems, respectively named the school-based model and the employer-led model.

The School-Based Model

In a school-based skill development system, the state serves as a major provider of initial vocational skills. The typical mechanism that compels employers to pay for VET is state-mandated taxes and special funds. Although taxpayers foot the bill, decisions regarding how to use the money—who to subsidize, what programs to fund, and levels of student enrollment—are made by the state. In this system, a significant proportion of general skill training transpires through public education subsidized by the state.

Systems that rely on the formal public education system to give the workforce initial vocational skills are often supplemented by extensive workplace-based continuing training that offers specific skills. The state often encourages this approach because it ensures that employers have a say in vocational training processes, and the approach prevents the school-based system from being disconnected from industry and overly focusing on general skills. A commonly employed state strategy is to subsidize firm-based training. Money collected through taxation comprises special funds that employers can obtain for training purposes. In the United States, the

Department of Labor funds two major federal-level training programs: Registered Apprentices and Job Corps. The Job Corps focuses on young trainees from economically disadvantaged families, and the Registered Apprentices funds training programs provided by employers or groups of employers—sometimes with the help of labor unions—that imitate the much-admired German apprenticeship system (Stone III & Lewis, 2010). At the local level, states have experimented with similar efforts (Osterman & Batt, 1993). Federal and state governments provide grants for employer-based training programs, and individual employers decide what programs to offer and what modules to include in the training. Scholars estimate that in the United States these state-subsidized training programs have not been very influential (Parker & Rogers, 1999; Stone III & Lewis, 2010).

More successful school-based national skill formation systems that feature mutually supplementary public and private trainings are seen in both Singapore and Japan. In the Singaporean model, there is greater state intervention than in the United States and Japan. Kuruvilla et al. (2002) ascribe the success of the post-war Singaporean economy to what they call a model of private-public sector collaboration in development. In this collaboration, the government ran a public education system that offers comprehensive vocational training, and it taxed the equivalent of 1% of the payroll of employees who earned less than $1500 per month. It then used these taxes to finance a special skill development fund. Firms that carried out training programs often recouped up to 80% of their contribution from this fund to cover their training costs.

During the early 1980s, the state also encouraged foreign-invested firms to set up training centers in order to secure technology and skill transfer to the domestic industry. To incentivize firms, the Singaporean government offered participating firms preferential investment and market protection opportunities and gave these firms preferred access to hire the graduates of the training institutes. The government later expanded this project to include other national governments, extending preferential opportunities to all firms from these countries. This model successfully encourages foreign firms to invest in skill development—ensuring that up-to-date technologies and skills are transferred to the Singaporean workforce—and secures a sufficiently large high-skilled labor force for the local establishments of these foreign firms. The state-organized Council for Professional and Technical Education was created in 1979 to coordinate training in both vocational schools and employer-funded training institutes (Kuruvilla & Chua, 2000). In Singapore, civil society actors do not play a significant role in this process.

In Japan, firm-based continuing training is an indispensable component of the skill development system that has played a pivotal role in the post-war economic miracle. Employers offer comprehensive and continuous vocational training throughout employees' careers, motivated by mutual commitments between employers and employees, who both expect lifetime employment. This cultural embedding within the Japanese employment relations system ensures that workers are deeply integrated into their firms' operations and growth. The hallmark of the lifetime employment system is its internal labor markets, providing extensive on-the-job training, regular bonus payments, and job security in exchange for unwavering work commitment. This model achieves two key objectives: first, it ensures that both employers and workers co-invest in the development of specific human capital; second, it mitigates the risk of free-riders in the external labor market. Furthermore, this arrangement guarantees that employers have ongoing input into the skill development process throughout their workers' careers, ensuring they benefit from their investment in training (Hashimoto, 1979; Hashimoto & Yu, 1980).

In school-based national models such as those seen in Singapore, state-funded public education provides comprehensive general skill training, supplemented by extensive employer-offered training focusing on specific skills. The Singaporean government adopts an interventionist approach, engaging employers through funding employer-offered training programs and encouraging foreign firms to establish training centers. Cost-sharing is achieved through comprehensive taxation policies and preferential trainee recruitment practices. In contrast, the Japanese system is rooted in unique employment relations characterized by internal labor markets and lifetime employment, ensuring strong work commitment and effective labor-capital collaboration. Both systems incentivize employers to invest in training, ensuring fair returns on these investments.

The Employer-Led Model

In relatively decentralized employer-led or mixed systems, such as those found in Germany and the United Kingdom, employers actively participate in pre-employment skill formation through innovative institutional arrangements designed to coordinate cost-sharing among all stakeholders. In Germany's dual system, the state has established an institutional framework for apprenticeship training, offering various resources and incentives to engage private players. Since the late 1960s, the German corporatist government has embraced the craftsmanship tradition of apprenticeship training, enshrining it into law. Under this system, youths receive training both in workplaces and schools under contracts negotiated between employer associations and unions, covering training programs lasting three to three and a half years (Culpepper, 1999a). The state establishes a tripartite institutional framework that allows private governance to autonomously coordinate skill development agendas, avoiding state centralization's limitations (Streeck, 1987).

Employer participation is crucial because the training process needs industry input and partial funding to ensure the latest technology and knowledge are transferred to schools and trainees gain real workplace experience. In Germany, this is facilitated through employer associations and chambers, forming a corporatist social partnership that lays the foundation for private governance of skill development. A tripartite framework facilitates employer engagement—even when such involvement may not be economically advantageous in the short term (Finegold & Wagner, 2002)—by coordinating the interests of employers and workers at the industry level. Key roles are divided among social partners: employers are approved, and training processes are monitored and evaluated by industry, commerce, and trade chambers. The state, via the Federal Institute for Vocational Training, provides expertise for designing apprenticeship programs. Collective labor plays a unique role in the German system, particularly in negotiating apprenticeship program contents at the industry level between employer associations and unions. Works councils at individual firms have rights to co-determine workplace-specific issues, helping prevent deviations from training agendas and protecting apprentices from abuse (Culpepper, 1999a).

These private governance arrangements ensure coordinated training processes at the industry level, preventing excessive focus on non-transferable skills. Even without effective tripartite frameworks, private governance can significantly influence skill standards setting, as demonstrated in the United Kingdom. Here, where unions play a minor role, training content is determined through negotiations among employers, with the state subsidizing training programs but leaving standard-setting and qualification to licensed private actors. Sectoral Skills Councils (SSCs), founded by employers at each national industry level, establish national vocational standards and issue certificates through approved organizations, ensuring widely transferable skills recognized by specific industries (Delebarre, 2016).

In summary, civil society actors, especially employer associations, and negotiation processes in skill standards making are essential for the success and sustainability of employer-led national systems. These processes help secure industry-level skill transferability, preventing individual employers from dominating training and over-focusing on specific skills. The state engages employers either directly through funding, as in the United Kingdom, or indirectly through private governance mechanisms involving employer associations and labor organizations, pressuring firms to invest in training.

By focusing on ideal formats of these national models, I identify necessary institutional preconditions for successful skill development and understand the rationales underlying these preconditions. These insights illustrate the needs for cost-sharing in skill development and show how successful national models actively engage employers and achieve skill transferability at the industry level. Drawing from international experiences and these ideal models, a theoretical framework can be established to evaluate the success of reform efforts in skills development.

How Do We Gauge the Success of National Skill Development Efforts: A Theoretical Model

Theoretically, a VET system must address three key problems to deliver desirable skills development outcomes: the underinvestment problem, skills mismatches, and imbalanced skills training. I delineate these problems and potential institutional resolutions for tackling them in this section, drawing on the experience of exemplar national VET models in other countries.

Problem I: Underinvestment in Skills Development

According to institutional theory, the investment of relevant parties in skills development based on their own rational choice tends to fall below the socially optimal level for three major reasons. First, skills are inherently public goods, thus, self-interested, utilitarian market players do not have sufficient incentive to contribute given the potential presence of free-riders. That is, human capital investment is a sunk cost for firms because other firms can potentially benefit from that investment by poaching their labor with slightly higher wages. As a result, rational individual firms are trapped in a prisoner's dilemma. Since workers can quit whenever they choose under the free employee termination regime, there is underinvestment in skills at the societal level, rendering everyone worse off (Streeck, 1989).

Second, the skills formation process has positive externalities: individuals and firms benefit from each other's human capital investment even though they may not directly use those skills. This is illustrated as a high-skill equilibrium scenario in which VET institutions, industrial relations systems, and production organizations, among others, can jointly promote an institutional environment that better appreciates and rewards skills. Workers end up better off, and the industry acquires a higher level of workforce productivity and additional bonuses as a result of the economic boom (Finegold & Soskice, 1988). However, an individual's level of investment in human capital is often based on his/her own cost/benefit calculation, which tends to be below the socially optimal level.

Finally, young workers at the start of their careers are unlikely to have sufficient money or motivation to invest in human capital that will only benefit them later in their careers because of the uncertainty associated with such an investment. Workers simply cannot accurately predict what skills will be needed in the future, thereby leading to insufficient self-financing of skills. Overall, human capital investment based solely on market logic tends to insufficiently fund skills. Lucas, for example, proposed a model that suggested that the United States economy would have reached the socially optimal level in the 1980s if it had invested almost three times as much in human capital as it had done (Lucas, 1988).

In institutional theory, therefore, skills are deemed as public goods. Streeck (1989) summarizes the market failure in the provision of sufficient skill training as follows:

> ...successful self-interested, utilitarian behavior in market environments requires the presence of collective resources, common values, and shared expectations that rationally acting individuals cannot normally generate, protect, or restore even if they fully recognize their vital importance. This is because such resources are in significant respects 'collective goods' which cannot be privately appropriated and to whose generation rational capitalist actors have therefore no, or no sufficient, incentives to contribute. As a consequence, the unbridled pursuit of self-regarding interests results in suboptimal outcomes not just for the community at large but also for economically rational individuals themselves.

As the market fails to result in the best outcomes for society, a type of collective governance is therefore indispensable for tackling the underinvestment problem (Olson, 1965). The key policy issue here is that the state should intervene to promote fair cost-sharing among workers, employers, and the state. International experience noted above has shown that exemplar national skills formation systems adopt one of three interventionist strategies to bring everybody on board. The first and most straightforward strategy is state subsidies for VET. While, globally, governments subsidize education and training, this approach is particularly favored by liberal market economies as it entails less direct intervention than other solutions. For example, in the United States, the VET system, which includes both schools and workplace-based training programs, such as apprenticeships, receives state funding, allowing the government to influence the training process in a way that is less interventionist and more market-friendly (Stone III & Lewis, 2010). In the United Kingdom, the state subsidizes training programs, but leaves standards setting, program development, and qualifications to licensed private actors (Delebarre, 2016). In China, a similar strategy has been adopted in some parts of the VET system; the state covers the tuition fees of students with an agricultural *hukou* or those who are enrolled in an agriculture-related program in both public and private vocational schools.

The second strategy involves the state obliging employers and individuals to share the costs of training. Although enforcing a direct state mandate on employers and individuals to pay for certain training in major countries is rare, the state may do so indirectly through taxation. In postwar Singapore, as said, employers were required to contribute the equivalent of 1% of the payroll of employees earning less than $1500 per month to a special skills development fund, which was granted back to firms that carry out training programs (Kuruvilla et al., 2002). The state may also make individual trainees bear part of the costs by (a) allowing lower apprentice wages than the legal minimum standards for ordinary workers, (b) allowing employment contracts stipulating mandatory post-training severance periods and liquidated damages on breaching such agreements, and (c) encouraging long-term employment relationships and discouraging labor poaching under legislative arrangements.

The third strategy is for states to make institutional arrangements in support of private governance. The German apprenticeship system, as noted, exemplifies a social partnership model of skills development with a clear division of labor among stakeholders. The German corporatist governance of the dualist system features training young people in both workplaces and schools under contracts negotiated between employer associations and unions. Employers are approved, and training processes are monitored and eventually evaluated by chambers of industry and commerce, and of trades. In workplaces, works councils play a role in co-determining firm-specific issues regarding training. Hence, the state diffuses authority to this private governance at the workplace and industry level and only serves to provide advice on designing apprenticeship programs. This institutional structure has successfully resolved the underinvestment problem by encouraging cost-sharing among a) the state that pays for vocational school operation, b) firms that pay for workplace training, and c) apprentices that contribute by working at wage rates lower than that of ordinary workers (Culpepper, 1999a, 3–5).

Overall, the three strategies for state intervention described above are indispensable for addressing the underinvestment problem in skills development caused by the free-rider problem, positive externalities, and uncertainty.

Problem II:Skills Mismatches

State subsidies or mandates may be able to encourage a wide range of employers to participate in an apprenticeship program for one cycle; however, their willingness to commit in the long run hinges on whether the program actually caters to their needs. The second crucial issue in skills development is thus the potential skills mismatches between VET processes and industrial needs. Especially in a country such as China, where regional variation is huge, curricula development should have the flexibility to accommodate regional and industrial specificities. Therefore, although state intervention can enable the coordination of firms and potentially overcome the underinvestment problem, employers may still be reluctant to participate when considerable skills mismatches exist.

The skills mismatches concern is derived from the French regulation school that is based on the Marxist theoretical proposition that the "economic base" or mode of production defines the limits of variation of a superstructure (Marx [1859] 1978)—this is understood as a configuration of institutional arrangements that the ruling class uses to govern society. It advances the notion of modes of regulation as a systematic mechanism for adjusting existing institutional forms and modes of production toward a coherent social-economic system (Boyer, 1990; Leborgne & Lipietz,

1988). In other words, at the core of regulation theory is the compatibility between institutions and industrial society.

This theory, if applied to skills development, suggests that institutional arrangements for skills development must be adaptable to the latest technologies and process innovations adopted in industrial organizations. Otherwise, the problem of skills mismatches will be present. Fordist mass production that once dominated the manufacturing industry in the United States, for example, was associated with many features of the American skills formation system, including the extensive use of general and standardized skills, limited firm-specific training, and a low level of workforce skills in general. This consistency lasted until the 1970s when the Japanese lean production system started to challenge the Fordist model in the North American market, and workers' skillsets became mismatched with the industrial needs. Firms have since integrated post-Fordist components into their production processes, including teamwork and worker participation. These changes have provided workers with a higher level and wider array of skills that have enabled problem-solving and multi-tasking in a more flexible production system (Green, 1992).

The diversified quality production model in post-war Germany, which is different from Fordism or the just-in-time production system, featured incremental customization that allows firms to afford the relatively costly human capital investment in young workers by hosting apprentices (Sorge & Streeck, 1988; Streeck, 1992). In turn, these workers equipped with specialized skills supported the diversified quality production model and sustained the comparative advantage of German products in the global market (Culpepper, 1999b).

These examples indicate that skills formation processes should engage firms and accommodate the types of skills needed by the industry. A certain diffusion of authority and school-firm collaboration is thus indispensable for bridging the skills mismatches. At the micro level, institutional arrangements for VET should grant a voice to the industry in skills standards setting, curriculum development, and qualifications as well as encourage firm participation in the pre-employment skills formation process (Culpepper, 1999a, 4). As noted above, the German VET system has been rooted in a corporatist regime with a strong social partnership tradition that has helped coordinate the training process among schools and industry as well as among employers within an industry.

The governments in East Asian developmental states tend to assume a larger role than those in European corporatist countries and are thus more comparable to China. Successful post-war industrialization in Singapore, Taiwan, Hong Kong, and South Korea was ascribed to the state's top-down efforts to not only facilitate the development of a skills formation system, but also proactively shape and coordinate demand for and the supply of skills in the market in the absence of a strong private governance tradition in these (once) authoritarian regimes (Ashton et al., 1999). However, this developmental state model of VET sometimes still relies on school-firm collaboration. In Singapore, the state-organized training agenda has encouraged foreign-invested firms to establish training institutes since the 1970s, allowing employers to enjoy preferential business investment and market protection policies as well as priority over other firms in hiring graduates from those institutes. The government later expanded this project to collaborate with other national governments and accordingly offered such preferential policies to all firms in those countries (Kuruvilla et al., 2002). By encouraging foreign firms to invest in skills development, this model has ensured that up-to-date technologies and skills are transferred to the Singaporean workforce and sufficient high-skilled workers are available to these firms. In conclusion, integrating industry needs through school-firm partnerships is important for tackling the skills mismatch problem and enabling a successful skills development system.

Problem III: Imbalanced Skills Training

While firm participation can resolve the problem of skills mismatches, it should not be the only goal of skills development. Ultimately, we aspire to have a balanced delivery of general and specific skills. Employer-sponsored training programs tend to excessively focus on specific and immediate skill needs of individual firms, and may therefore fail to fulfill China's long-term upskilling agenda. As said, human capital theory distinguishes general skills that equally benefit every firm from specific skills that only benefit a certain firm (or group of firms) (Becker, 1993). However, Stevens defined most skills as being transferable skills on which other firms can capitalize, but not to the extent that—assuming they are subject to perfect market competition—"the wage is driven up to the marginal product" (Stevens, 1999).

Employers have strong incentives to focus their efforts on developing specific skills to avoid free-riders: workers equipped with more specific skills are less attractive in the labor market, and thus, less likely to be poached by other firms than those with more general skills, assuming other factors to be constant. Therefore, an employer who is invited to participate in curriculum development undertakes a training process that risks being dominated by the specific skill needs of the firm at work.

This poses a great challenge not only to individual trainees, but also to the economic development in general. Overly focusing on firm-specific skills undermines workers' employability in the labor market and limits their long-term career development (Locke et al., 1995). A workforce with a skill structure biased toward the specific needs of individual firms undermines labor market mobility and increases retraining costs for new workers. The absence of a balanced plan in skills development at the societal level will thus eventually harm the economy.

The solution here is to establish mechanisms that can coordinate the interests of employers—preferably at the industry level—in skills standards setting, curriculum development, and qualification and process monitoring to balance general and firm-specific skills and achieve sufficient skills transferability. For institute-based training, a skill credentialing system should be established that focuses on occupation-related standards setting and certification for transferable skills at the industry level. These standards could be used to guide the development of training programs and curricula to ensure that essential occupation-specific skills are covered in training. In the United Kingdom, for instance, sectoral skills councils establish national vocational standards, which are then used to guide formal institute-based training programs, and issue certificates from approved organizations to attest to receivers' employability for a particular occupation (Delebarre, 2016).

For employer-sponsored training, such as apprenticeship programs, worker voice and workplace monitoring mechanisms are necessary to overcome the imbalance problem and avoid misallocating apprentices (as "labor agencies"). Civil society governance, particularly labor unions, play a role here, as they are incentivized to strengthen the employability of member workers over time and across firms (Olney, 1994). In Germany, as noted earlier, training content is negotiated between employer associations and unions at the industry level. In the workplace, works councils have the right to co-determine issues that are firm-specific and monitor the process daily. Finally, chambers of industry and commerce evaluate apprentices as a quality control mechanism. Therefore, these civil society actors help prevent the deviation of the training process from its stipulated agendas and from being overly firm-specific (Streeck, 1987; Culpepper, 1999a).

In conclusion, a successful skills development system involves not only training institutions, but also extensive collaboration among the state, civil society actors, and employers to address the three core challenges outlined above. This discussion provides a set of criteria and indicators particularly useful for evaluating school-firm collaboration and apprenticeship reforms—topics that will be analyzed in detail in Chaps. 4 and 5.

Studying VET in China

This study is based on 25 vocational schools in China's Guangdong Province, selected during a two-month preliminary research phase in 2014–2015, followed by one year of intensive fieldwork conducted in 2015–2016. I selected Guangdong as an information-rich case (Patton, 1990: 181), given its dynamic economic development and the diversity of its VET landscape. Widely recognized as the heart of the Pearl River Delta, Guangdong has developed the most vibrant regional economy in China over the past three decades, hosting the country's most advanced manufacturing and service industries. If Guangdong's rise symbolizes the achievements of China's 40-year economic reform, it also exemplifies the most pressing challenges currently constraining the "Made in China" model. In particular, the province's rapid economic growth is hindered by a severe shortage of skilled labor, rising labor costs, and increasing labor unrest. Addressing these issues demands industrial restructuring and upgrading—and, consequently, an upgrade in workforce skills. Given their historical effectiveness in implementing innovative economic reforms, Guangdong's state officials and employers are likely to spearhead efforts to resolve the skilled labor deficit through VET reform. We should therefore expect to observe the most dynamic forms of school-firm collaboration and the most creative reform initiatives in this province.

Moreover, Guangdong is China's largest provider of vocational education. By 2014—the year I began my field research—the province hosted 574 vocational schools and 243 technical schools, accounting for 5.5% and 8.6% of the national totals, respectively. That year, 2.7 million students were enrolled in these institutions, representing approximately 9.8% of the national enrollment—more than any other province (more data on Guangdong VET is provided in Chap. 2). In addition, Guangdong exhibits significant regional variation. The Pearl River Delta region—comprising cities such as Guangzhou, Shenzhen, Foshan, and Dongguan—features the country's most developed market economies and highest-performing vocational schools. In contrast, peripheral areas, including Qingyuan, Zhanjiang, and Zhaoqing, resemble the less developed provinces of China's interior. While Guangdong is not a prototypical case, it is highly representative of the diverse conditions within China's VET system, making it an ideal, information-rich site for analysis.

Table 1.1 presents the 25 schools examined in this study. The sample includes both vocational and technical schools. Within each category, I selected at least one high-performing, one mid-tier, and one lower-tier institution, based on criteria such as training effectiveness, funding sources, graduates' skill levels and employment prospects, and official and social recognition. This stratified approach enabled comparative analysis across school types and performance levels, while also identifying common patterns and systemic challenges. The schools are geographically dispersed across Guangdong, ranging from highly industrialized cities like Guangzhou, Shenzhen, Dongguan, Foshan, and Zhuhai to less developed regions such as Qingyuan, Zhanjiang, and Zhaoqing. Notably, high-performing schools are not exclusively located in the most advanced cities, although local government support does significantly influence institutional development, as shown in Chap. 2. Guangzhou, as the provincial capital and educational hub, hosts the largest number of schools, including many of the top-tier institutions.

Data

This study draws on three types of data collected during field research. First, I conducted 322 semi-structured interviews with various stakeholders in the VET system. The full interview list is provided in Appendix B. The 200 in-depth interviews with school principals, administrators, teachers, students, and parents from the 25 sampled schools constitute the primary source of insight into the daily operations of Chinese vocational institutions. Interviews with officials from the Ministry of Education (MOE) and the Ministry of Human Resources and Social Security (MOHRSS), as well as their local branches, provided a macro-level understanding of the VET landscape in both Guangdong and China more broadly. These conversations also helped identify and connect me with representative school cases across different tiers. As the state is a central stakeholder in the VET system, its role was explored through direct engagement. Given the extensive scholarly attention to school-firm collaboration as a key mechanism in VET, I also interviewed managers, union

Table 1.1 Distribution of schools studied in Guangdong

	Developed Areas					*Less Developed Areas*			
	Guangzhou	*Shenzhen*	*Dongguan*	*Foshan*	*Zhuhai*	*Qingyuan*	*Zhanjiang*	*Conghua*	*Zhaoqing*
High-end	GDPST (MOE) I GDIP (MOE) I GZPYP (MOE) I BYTCBT (MOHRSS) I GDMTC (MOHRSS) I	SZP (MOE) I		SDLVTS (MOE) III		QYP (MOE) III			
Mid-range	GDCP (MOE) II GZHSC (MOE) II GZRP (MOE) II GDCP (MOE) II GDMEC (MOE) II GZVSLI (MOE) III GDLNCIC (MOHRSS) I GZTC (MOHRSS) III		DGEES (MOE) III	SDVTS (MOE) III			ZJHS (MOE) II		ZQMC (MOE) II
Low-end	GDVCPT (MOE) II GZIT (MOE) III		Anonymous (MOE) IV		ZHHS (MOE) II			GZCHTS (MOHRSS) III	

Note: I, II, III, and IV correspond to skill development models in Table2.4 in Chap. 2. See the interview list in Appendix B for full school names representatives, and workers/trainees from 21 firms to assess their perspectives and involvement. Additionally, I spoke with officials from 12 other relevant organizations, including local trade unions, think tanks, NGOs, employer associations, and academic institutions, which provided valuable supplementary context. Some interviews were conducted in group settings—a common practice in China—while others were individual. On average, each interview lasted two hours. These in-depth conversations form the core of my empirical data, offering nuanced understanding of the VET system's functioning and challenges.

Second, to explore the linkages between vocational schools and other stakeholders, I conducted extensive participant and non-participant observation of their ongoing interactions. This was achieved in two main ways. First, I was granted access by government officials and school leaders to attend a wide range of meetings—public and private—as well as informal conversations and social dinners, which are a key mode of relationship-building in China. Second, I actively participated in two VET reform initiatives in Guangdong. From late 2015 to early 2016, I served as an external consultant for a curriculum reform project in Guangdong's beauty industry. In April 2016, I was invited to act as a translator for a team of British consultants supporting the MOE's apprenticeship reform pilot. These participatory roles provided rare insider access to the strategic processes through which VET practitioners align institutional operations with stakeholder interests and reform initiatives.

Finally, I collected a substantial body of secondary data to complement the primary sources. This includes national and local policy documents, internal reports from government agencies, schools, and firms, as well as internet-based materials such as institutional profiles, news reports, and public announcements. These materials not only informed the design of targeted interview questions but also helped fill gaps in the narrative that could not be fully reconstructed through interviews alone. Together, the triangulation of interviews, observation, and documentary evidence ensures a robust and comprehensive analysis of vocational education and training in contemporary China.

An Institutional Dilemma in China's VET System

My central argument in this book is that, in the absence of favorable institutional preconditions—such as an interventionist state, long-term employment relations, internal labor markets, or an effective private governance framework supported by proactive employer associations—the Chinese state has faced a profound dilemma over the past decade. This dilemma arises from its attempt to transform the national skill formation system from a centralized, school-based model into a more decentralized system that encourages, and in some cases even empowers, employers to take a leading role in vocational training.

As existing research suggests, a school-based model of VET can avoid an overemphasis on general skills provided that employers are actively involved in pre-employment or on-the-job continuing training and that such training complements the foundational skills delivered by public vocational institutions. However, scholars have consistently found that firm-based training remains extremely limited in China. When it does occur, it is largely confined to foreign-owned firms, joint ventures, and occasionally large state-owned enterprises (SOEs) and tends to target managerial and high-skilled personnel (Child, 1994; Cooke, 2005; Warner, 1993; Wang, 2014; Xiao & Tsang, 2004). Moreover, the state has largely failed to establish systematic mechanisms to incentivize employer participation in skill development (Cooke, 2005; Lai & Lo, 2006).

A key reason employers have remained passive is the absence of stable, long-term employment relationships in Chinese firms. With the dismantling of internal labor markets in the state sector and the informalization of employment in the rapidly expanding private sector during economic reforms, workforce turnover has remained high across most industries. In response to growing employment instability, the Labor Contract Law sought to promote long-term employment by imposing restrictions on arbitrary dismissal (Gallagher & Dong, 2011). Yet the law offers employers minimal protection for their training investments. Article 22 permits employers to claim liquidated damages from employees who resign after receiving specialized technical training, but only if a service period agreement has been signed in advance. Even then, damages are capped at the unamortized portion of training costs. In practice, most employers are reluctant to formalize such agreements for every training activity. This limited legal safeguard discourages firms from investing in comprehensive, firm-specific continuing training. During my fieldwork, numerous employers voiced frustration over this institutional shortcoming.

To address the lack of workplace-based training, the government has pursued a strategy of decentralization over recent decades, encouraging employers to engage in pre-employment training—specifically, the school-based VET process—through knowledge and technology transfer. Lai and

Lo (2006), who documented state-led decentralization efforts since the mid-1980s, concluded that these reforms had largely failed by the early 2000s. My research, however, indicates a significant shift since then: the state has now largely withdrawn from the day-to-day management of vocational schools, granting them near-total autonomy in designing and implementing training programs, while actively encouraging employer input into pre-employment training. As a result, the long-standing misalignment between vocational education and industry needs—highlighted by scholars two decades ago (Cooke, 2005; Durden & Yang, 2006; Lai & Lo, 2006)—has been substantially reduced.

Nevertheless, reform has swung to the opposite extreme. Most employer-involved training programs now focus narrowly on the immediate, specific needs of collaborating firms. The state has neither established widely recognized skill standards nor incentivized employers to coordinate training agendas across industries. In effect, the decentralization process in China has proceeded in a highly fragmented manner: individual firms have assumed control over training curricula previously managed by schools, but without any mechanism for cross-program coordination. Drawing on Western experiences, this problem of *over*-decentralization might have been mitigated by either robust private governance or a credible, industry-recognized skill certification system. However, in post-socialist China, civil society actors—particularly employer associations and labor unions—are state-supervised and lack both the autonomy and motivation to coordinate workforce skill development. Private governance, therefore, remains largely absent.

Furthermore, the national skill certification system—the National Vocational Qualification System (NVQS), administered by the MOHRSS—has been marginalized within both training processes and hiring decisions. This is because the NVQS fails to take into account a critical feature of contemporary Chinese industry: regional disparities in economic development. Different regions are at varying stages of industrial maturity, meaning that even within the same occupation, skill requirements differ significantly across localities. A uniform, top-down credentialing framework is thus ill-suited to China's diverse industrial landscape.

I conclude that the current skill development system in China is disorganized and fragmented. Individual employers—rather than coordinated employer groups—have replaced the state as the dominant force in shaping vocational training. Yet these employers have failed to institutionalize any mechanism for industry-level coordination. The result is systemic disorder, driven primarily by two interrelated factors: (1) the absence of autonomous employer associations capable of formulating coherent, cross-firm skill development strategies and (2) the state's failure to enforce meaningful skill standards at both regional and national levels. This has left training programs isolated, uncoordinated, and overly tailored to individual firms.

While this system effectively addresses the immediate and specific skill demands of employers—and thus helps to partially relieve the secondary labor market failure discussed earlier—it systematically underemphasizes general and transferable skills, which are equally vital for long-term workforce adaptability and employability. I argue, therefore, that China's VET system is caught in a *skill formation dilemma*: in the absence of effective private governance to reconcile the diverse interests of employers, neither a fully centralized school-based model nor a purely decentralized, employer-led model can produce optimal skill formation outcomes. This structural impasse poses a significant challenge to China's broader economic agenda of industrial upgrading and long-term upskilling.

Through an analysis of Guangdong Province—an information-rich case (Patton, 1990) in terms of reform dynamics and school-firm collaboration—I empirically examine this skill formation dilemma, the problems it generates, and recent state-led attempts to address the resulting disorder. I argue that VET policymakers and practitioners in Guangdong face a fundamental paradox. On one hand, decentralization has successfully integrated employer input into training, narrowing the gap between vocational education and real industry needs—a gap widely criticized in earlier scholarship. Driven by skilled labor shortages and labor market inefficiencies, employers have shifted from traditional in-house training to engaging with pre-employment skill development, seeking to secure firm-specific competencies and a flexible labor reserve.

On the other hand, this shift has introduced new challenges. Vocational schools now operate with considerable autonomy but lack binding industry-level skill standards. Employer involvement remains fragmented, making cross-industry coordination of skill transferability virtually nonexistent. The MOE and the MOHRSS have each launched independent apprenticeship reforms. The MOE continues to promote a decentralized, experimental model with minimal state oversight, allowing schools and firms to co-design training programs tailored to specific enterprise needs—once again prioritizing specificity over generality. In contrast, MOHRSS has adopted a top-down approach aimed at revitalizing the NVQS. Yet this effort remains disconnected from industry realities and fails to involve employers in the standard-setting process.

I contend that the root of the problem lies in the absence of effective civil society governance. Drawing on international experiences, effective VET systems require the active participation of employer associations and labor unions—not only to coordinate diverse employer interests but also to prevent training from becoming excessively narrow and firm-specific. Without such intermediary institutions, China's VET system risks remaining trapped in a cycle of disorganization, unable to balance immediate labor market responsiveness with long-term national skill development goals.

Stages of the Argument

The empirical core of this book unfolds across four interrelated chapters, each advancing a distinct analytical task while collectively building toward the central argument: that the evolution of China's VET system since the reform era can be understood as a three-stage process—decentralization, collaboration, and institutionalization—shaped by the interplay of institutional configurations, organizational agency, and market pressures.

The inquiry begins in Chap. 2 by establishing the historical and institutional context of the state's workforce upskilling agenda, particularly in the wake of the 2015 policy push to revitalize VET. Moving beyond reductive portrayals of vocational schools as mere labor agencies, this chapter develops a more granular typology of school-level practices. Drawing on fieldwork in a diverse sample of 25 vocational schools in Guangdong, it demonstrates how the decentralization of educational authority has led to substantial variation in training effectiveness, funding mechanisms, and student outcomes. Four distinct skill development models emerge: high-performance, industry-focused, local market-oriented, and labor agency. The persistence of labor agency practices is shown to be largely confined to under-resourced private institutions. The chapter argues that this variation is systematically shaped by a combination of institutional conditions—state support and local industrial structure—and organizational attributes—school legacy and ownership type. This analysis reveals the decentralized and heterogeneous landscape of vocational education, thus laying important foundation for the school-firm collaboration and the apprenticeship reforms examined in the subsequent chapters.

Chapter 3 shifts the focus to employer behavior, examining how firms respond to skilled labor shortages through strategic collaboration with vocational schools. Challenging the prevailing view that school-firm ties in China are weak or instrumental, the chapter identifies a widespread practice of "training for a targeted brand," wherein employers partner with schools to co-design training programs that embed firm-specific skills into formal curricula. This arrangement allows firms to externalize part of their internal training functions, mitigating labor market failures while securing a pipeline of pre-trained workers. In turn, schools gain access to advanced technologies and industry knowledge, enhancing their pedagogical offerings. The chapter theorizes this phenomenon as a form of institutional adaptation, emerging precisely because of the decentralized VET system and acute labor shortages. It underscores the strategic agency of firms in shaping skill formation outside their organizational boundaries.

Building on these findings, Chap. 4 evaluates the state's recent attempts to institutionalize and scale up such collaborations through two parallel apprenticeship reforms led by the MOE and the MOHRSS. Despite shared policy goals, the ministries adopt divergent institutional logics: MOHRSS pursues a top-down, centrally managed model emphasizing standardized quotas and general skills, while MOE adopts a bottom-up, collaborative approach that legitimizes and integrates existing "training for a targeted brand" initiatives. Applying the three-level analytical framework informed by institutional theory, the regulation school, and human capital theory established above, the chapter assesses the relative effectiveness of these models. While the MOE's collaborative strategy shows greater promise in addressing collective action problems and aligning training with industry needs, both reforms fall short of achieving a balanced provision of general and firm-specific skills—a shortfall attributable to the absence of strong civil society governance in China and proactive employer associations in particular. The chapter concludes that the Chinese state remains caught in a skill formation dilemma: without the emergence of effective private governance structures, its institutionalization efforts risk reproducing imbalances rather than resolving them.

Together, these chapters trace a dynamic trajectory—from decentralized variation to emergent collaboration and, finally, to contested institutionalization—offering a nuanced account of skill formation in contemporary China. The argument advanced is not merely descriptive but analytical: it reveals how structural constraints and strategic agency interact to shape the possibilities and limits of vocational education reform in a post-socialist developmental state.

CHAPTER 2

Decentralization: Withdrawal of the State and Varieties of Vocational Schools in China

In sharp contrast to its growing relevance in practice, scholarly understanding of the Chinese vocational education system remains inadequate and at times biased. A prevailing view in the literature casts Chinese vocational schools as quasi-labor agencies. Scholars argue that training processes in these institutions are largely ineffective and misaligned with market demands (Cooke, 2005; Durden & Yang, 2006; Lai & Lo, 2006), resulting in graduates who possess few marketable skills (Woronov, 2012, 2016; Song & Xu, 2024) and are subsequently dispatched to firms as seasonal, low-cost labor—often in exchange for unjustified commissions (Chan et al., 2015; Smith & Chan, 2015; Song, 2024; Su, 2010–2011). While these critiques highlight real problems, they require systematic updating. Much of this earlier research relies on piecemeal or secondary data, frequently collected before the turn of the century. The now-well-known labor agency practices were largely exposed through studies focused on Foxconn, raising concerns about industrial specificity. Moreover, it remains unclear whether these studies employed a rigorous or representative sampling strategy on the school side. These limitations suggest that existing accounts may overlook significant variation within the Chinese vocational education system.

This chapter argues that the decentralization reforms initiated in the mid-1980s have generated substantial diversity across vocational schools and that the "labor agency" depiction, while accurate for certain cases,

H. Zhang, *Upskilling the Chinese workforce?*,
https://doi.org/10.1007/978-981-95-2656-7_2

reflects only a subset of the broader institutional landscape. To capture and explain this variation, I draw on a diverse case sample of 25 vocational schools and 21 of their collaborative firms in Guangdong Province—an information-rich context (Patton, 1990) given the region's wide range of vocational education arrangements. My analysis identifies four distinct patterns of skill development: the high-performance model, the industry-focused model, the local market-oriented model, and the labor agency model. Schools operating under these different models exhibit varying degrees of training effectiveness, rely on different funding sources, and produce graduates with distinct skill levels and labor market scopes. This typology enriches our understanding of the Chinese vocational education system by revealing the considerable heterogeneity among vocational schools.

I account for these divergent development paths by linking them to four key institutional and organizational factors: state support and strategic orientation, local industrial structures, a school's institutional legacies, and its ownership. As an institutional backdrop, I first outline the structure of China's vocational education system and the diffusion of authority from the central government to individual schools—a process rooted in the decentralization reform. I then present the taxonomy of skill development models and examine the forces that shape them.

China's Vocational Education System

Figure 2.1 illustrates the structure of China's education system by age of enrolment. At the pre-employment stage, the vocational education system comprises two primary tracks: vocational schools and colleges administered by the MOE, and technical schools and colleges overseen by the MOHRSS. Students aged 15 to 18 who choose vocational pathways typically enter either a *zhongzhi* (中职) or a *zhongji* (中技), both of which confer qualifications equivalent to a general senior high school diploma. After graduation, they may enter the labor market directly or pursue further education at a *gaozhi* (高职), a *gaoji* (高技), or a *jishi xueyuan* (技师学院). Some vocational college graduates can subsequently enroll in two-year *zhuan sheng ben* (专升本) programs to obtain bachelor's degrees. These institutions and their associated pre-employment training programs constitute the core of China's vocational education system.

A defining feature of this system is its dual administration under the MOHRSS and the MOE. Table 2.1 compares the number of schools,

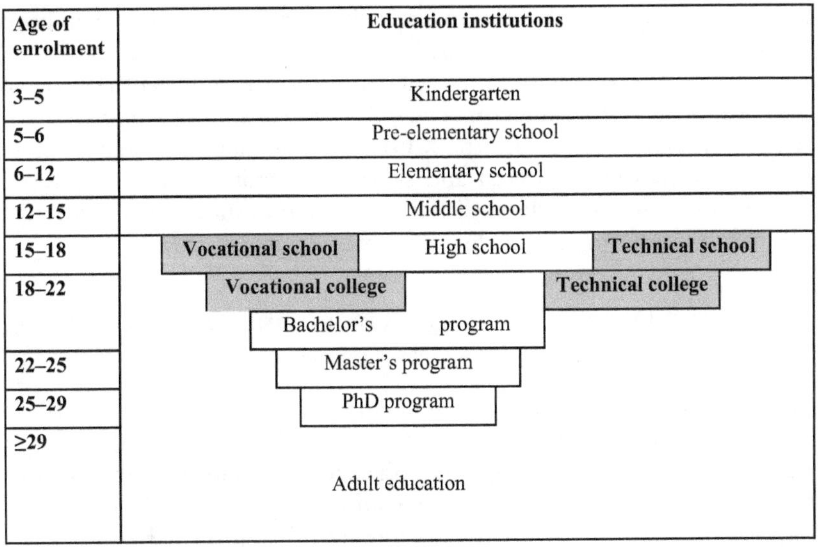

Fig. 2.1 China's education system by age of enrollment

Table 2.1 Technical and vocational school in China and Guangdong, 2014 and 2023

	Year	*Schools*	*Teachers*	*Students*
MOHRSS—Technical schools and colleges	2014	2818 (243 in Guangdong)	194,631 (20,840 in Guangdong)	3,389,696 (622,614 in Guangdong)
	2023	2468 (148 in Guangdong)	N/A	4,395,000 (658,700 in Guangdong)
MOE—Vocational schools and colleges	2014	10,387 (574 in Guangdong)	1,102,082 (N/A)	24,229,127 (2,078,207 in Guangdong)
	2023	8632 (464 in Guangdong)	1,419,426 (101,500 in Guangdong)	30,063,142 (2,160,000 in Guangdong)

Data Sources: 2014 data: China Statistical Yearbook 2015; Guangdong Education Bulletin 2014–15; 2023 data: China Statistical Yearbook 2024; MOHRSS Statistical Yearbooks 2023; Guangdong Education & HRSS Bureau Reports 2023 teachers, and enrolled students in institutions under each ministry in both China and Guangdong, as of 2014—when this research began—and in the present. As shown, both ministries administer a significant portion of the vocational education sector. The MOE oversees more schools, teachers, and students overall, but the MOHRSS maintains a substantial presence, particularly in technical and skills-based training. This bifurcated governance structure reflects differing policy priorities and institutional cultures, contributing to the diversity observed across schools.

The Decentralization Reform and Diffusion of the State Authority

The Chinese vocational education system has long been criticized for failing to meet the skill requirements of industry (Cooke, 2005; Durden & Yang, 2006). Curricula and training practices were widely seen as disconnected from workplace needs, lacking sufficient practical components. In response, China launched a major decentralization reform in the mid-1980s (Lai & Lo, 2006), marking a shift away from centralized control. The state attempted to withdraw from the daily operations of vocational schools, granting them autonomy to design and implement their own training agendas, and encouraging employers to contribute to pre-employment education through knowledge and technology transfer.

Nevertheless, by the early 2000s, the system remained largely disengaged from industry, and skills mismatches persisted (Cooke, 2005). Some programs were criticized for failing to equip students with useful competencies (Durden & Yang, 2006), instead functioning as labor intermediaries that placed student interns on mass production lines for questionable fees (Chan et al., 2015; Smith & Chan, 2015; Song, 2024). While some school-firm collaborations emerged, they were often based on personal networks and lacked institutional stability (Lai & Lo, 2006). This partial failure of decentralization prompted the Chinese state to initiate new reforms—examined in Chap. 4—modeled after the German apprenticeship system, aimed at strengthening industry involvement and upgrading workforce skills. This chapter aims to provide an updated overview of the Chinese VET system as it stood at the time of these reforms implemented in 2015.

During my preliminary fieldwork, one observation stood out as strikingly at odds with the image of the vocational education system portrayed in the literature: the extent of decentralization. As my research deepened,

I found that state authority has diffused significantly, with vocational schools now exercising near-complete autonomy in shaping their training processes. This counterintuitive reality—of a system in which the state has effectively retreated from direct oversight—became the empirical foundation for this study.

Table 2.2 summarizes the degree of authority diffusion in three key domains: the establishment of new specialties, student admissions, and curriculum development. In most cases, de facto decision-making occurs at the school level or below, with local governments retaining authority only over a limited set of issues, such as the NVQS, a small number of mandatory courses, and the formal approval of specialty programs. Even in these areas, schools frequently find ways to circumvent or negotiate state intervention, preserving considerable operational freedom.

This widespread autonomy has enabled schools to respond flexibly to local labor market conditions and industrial demands, but it has also contributed to uneven training quality and institutional divergence. The retreat of the state has not led to uniform outcomes; instead, it has created the conditions for multiple models of vocational education to emerge, shaped by local economic contexts, historical legacies, and organizational strategies. It is within this decentralized landscape that the four models identified in this chapter have taken root, reflecting both the opportunities and challenges of a system no longer centrally directed.

Table 2.2 Authority diffusion in the Chinese vocational education system

	MOE		*MOHRSS Technical Schools*
	Vocational Colleges	*Vocational Schools*	
Specialty establishment	State approval needed	School	School
"Direction"	School	School	School
Student admission	*Gaokao*—plus a significant proportion of school autonomous admission	School	School
Qualification	NVQS and national degree	NVQS and national degree	NVQS and school degree
Curriculum development	Department, and state mandatory courses	Department, and state mandatory courses	Department, and state mandatory courses
Textbooks	Teacher	Teacher	Teacher

Specialty Establishment

In the literature, scholars argue that the state tightly controls the establishment of new specialties in vocational schools through local government-organized specialty committees and national specialty indexes. By strictly adhering to the relevant specialty index, which is updated only every three to five years (Chen, 2015), such committees retain the authority to approve or reject a school's proposal to launch a new specialty, potentially leaving institutions with little room for maneuver (Lai & Lo, 2006). What I found in Guangdong's vocational schools, however, diverges significantly from this portrayal. Across all institutions in my study, school officials consistently expressed the belief that they could establish new specialties without difficulty, reflecting a reality of considerable institutional autonomy.

To understand this process, it is necessary to distinguish between two scenarios. The first pertains to the establishment of indexed specialties—those already listed in the official national or provincial catalogues. In interviews conducted with officials from the Guangdong Department of Education (GDDOE) and the Guangdong Department of Human Resources and Social Security (GDDOHRSS), who are responsible for overseeing school affairs, a clear pattern of administrative non-intervention emerged. As one official explained:

> With only a few exceptions, establishing specialties in Guangdong vocational schools is completely free from state control… Schools enjoy full autonomy and need not seek our approval, including for non-indexed specialties. National policies issued by the MOE require that non-indexed specialties be registered with the provincial educational department, but Guangdong is more advanced in this regard than the national standard. We believe even non-indexed specialties need not be registered, leaving autonomy entirely to the school. (Author: What about vocational colleges?) The situation is largely the same, except for one procedural difference—their specialty establishment must be recorded with us. But note, this is not approval! …Vocational education must align with local economic development. Schools should be able to adapt dynamically to local economic needs, particularly in specialty development and student enrollment. By granting them autonomy, schools can realize their potential to respond effectively to market demands. This is our guiding principle. (I: 5)

Similarly, an official from the technical education division under the MOHRSS elaborated on the rationale behind this hands-off approach:

> Technical schools historically originated as enterprise-run institutions, designed to train skilled workers directly for industrial firms. Their training processes must therefore remain closely aligned with firm requirements and emphasize practical skills. For this reason, the MOHRSS mandates that schools maintain strong connections with employers and the labor market… Our focus is on how to encourage firm participation in education—through program design, teaching capacity development, and curriculum formulation—so that training keeps pace with technological change. Government agencies do not intervene in these processes but allow schools and firms to collaborate freely. (Author: Are schools free to determine specialty establishment and student enrollment?) Yes, completely free! (I: 8)

The second scenario involves the establishment of ***non-indexed*** specialties—those not yet recognized in the official specialty list. This is arguably the more critical issue, given the rapid pace of market transformation and the inherent lag in bureaucratic index updates. If schools were restricted to only those specialties listed in the index—even under a relatively permissive approval regime—they would face significant constraints whenever the index failed to reflect emerging industries or skill demands.

As the interview evidence indicates, the Guangdong provincial government has granted schools greater latitude than the national framework requires when establishing non-indexed specialties. Yet even in the absence of such provincial flexibility, schools have developed institutional workarounds to navigate index limitations. A common strategy is to introduce a new training focus under the umbrella of an existing indexed specialty by designating it as a *zhuanye fangxiang* (专业方向), or “specialty direction.” This allows schools to offer curricula tailored to new industries while formally remaining within the boundaries of an approved specialty. Over time, such directions are often incorporated into official updates of the national specialty index. This practice, widely accepted across Chinese vocational institutions and universities, functions as a de facto probationary mechanism that enables experimental responses to market changes before formal institutionalization.

In the field of vocational education, for example, training programs related to robotics began appearing in schools during the early 2010s, coinciding with the rise of automation and smart manufacturing in China.

However, "Robotics" did not appear as an officially recognized specialty until the MOE updated its index in 2015. In the interim, schools circumvented the restriction by establishing "Robotics" as a *direction* under indexed specialties such as Computer Numerical Control, Mechatronics, Internet of Things, and Electronics Application. These directions included specific course sequences and practical training modules in industrial robotics, robot application and maintenance, and smart manufacturing systems. Because schools retain full control over curriculum design and training implementation—discussed in detail below—specialty review committees have limited capacity to monitor the substantive content of these new directions or assess how significantly they deviate from the parent specialty. This form of institutional adaptation enjoys tacit state approval.

An official from the Dongguan Bureau of Education illustrated the practice:

> Many schools avoid establishing a completely non-indexed specialty. Robotics, for instance, was not officially indexed until last year. But the Dongguan Science and Technology School offered this program under the name Mechatronics, with a specialty direction in Industrial Robots... In their recruitment materials, they prominently feature "Industrial Robots," because otherwise students would not understand what kind of work the program prepares them for. (I: 11)

This adaptive strategy enables schools to respond swiftly to transformative shifts in the labor market and to align their offerings with contemporary industrial trends, unencumbered by rigid administrative classifications. While the MOE did not formally recognize any robotics-related specialty until 2015, media reports indicated that by 2014, 9 vocational schools had already launched full specialties, and 120 had established specialty directions in robotics across China.[1] A parallel case occurred with 3D printing. Vocational schools introduced specialty directions into additive manufacturing well before 2015, when the subject was officially included in the national specialty index. Indeed, several schools in my sample were already offering robotics- and 3D printing-related training as early as 2011 and 2013, respectively.

[1] See "Jiqiren shidai laile, caokong rencai zaina?" (The age of robots has arrived, where are the talents who control them?). *China Youth Daily*, March 2, 2015, https://zqb.cyol.com/html/2015-03/02/nw.D110000zgqnb_20150302_1-10.htm. Accessed July 2, 2025.

Table 2.3 presents a comparative analysis of vocational school graduates across industrial sectors in Guangdong, contrasting data from 2012 and 2016. The shift in educational focus mirrors broader economic transformations in the province, particularly deindustrialization and the expansion of the service sector. Over this period, vocational schools reduced their emphasis on manufacturing and increased offerings in healthcare, social work, education, and the arts. The table also compares Guangdong's graduate distribution with the national average in 2014. Notably, the proportion of graduates entering the primary and secondary sectors—agriculture, manufacturing, and construction—is significantly lower in

Table 2.3 Graduates placement in Guangdong and the entire country across industries in selected years

	Guangdong 2012 (%)	*Guangdong 2014 (%)*	*National 2014 (%)*	*Guangdong 2016 (%)*
Agriculture	3.5	*5.7*	*10.1*	3.2
Natural Resources and Environment	0.3	*0.1*	*1.6*	0.3
Energy	0.3	*0.1*	*1.0*	0.2
Construction	1.3	*1.7*	*3.7*	2.2
Manufacturing	*15.7*	*12.9*	*18.2*	*13.2*
Petro-chemistry	0.3	*0.5*	*1.0*	0.4
Textile and Food	0.8	1.5	1.4	1.2
Transportation	5.5	6.0	7.4	6.5
IT	20.2	*18.6*	*16.2*	16.5
Healthcare	*9.9*	9.9	9.8	*11.5*
Leisure and Health	0.7	0.3	0.8	0.3
Finance, Business Management, and Commerce	22.1	*20.6*	*9.6*	21.5
Tourist Service	3.8	3.9	4.1	3.0
Arts	*3.6*	3.9	3.5	*4.5*
Sports and Fitness	0.4	0.7	0.7	0.3
Education	*6.0*	*8.3*	*6.2*	*9.4*
Legal Service	0.4	0.3	0.4	0.3
Public Affairs and Social Work	*2.3*	1.9	1.7	*2.8*
Others	3.0	3.3	2.4	2.8

(Data sources: 2012 and 2014 data from Reports on Chinese Vocational School Graduates' Employment Conditions 2006–2012 and 2014 by the MOE; 2016 data from the Guangdong Department of Education)

Guangdong than in most other provinces. Relative to the national average, Guangdong has produced a higher share of graduates in information technology, finance, management, and commerce, reflecting its long-standing emphasis on industrial upgrading. Moreover, as a resource-scarce coastal province, Guangdong enrolls far fewer students in energy and natural resource-related fields. As the data demonstrate—and irrespective of national specialty indexes—local authorities have permitted schools broad discretion in shaping their specialty portfolios in response to regional economic dynamics.

Student Admission and Qualification

With regard to student admission, one significant constraint facing applicants to vocational colleges lies beyond schools' control: the *gaokao* (高考), or National College Entrance Examination. This standardized test remains a central gatekeeper for access to higher education in China. However, for most other types of vocational institutions—such as technical schools and secondary vocational schools—local authorities have granted full autonomy in determining student recruitment. Despite the formal centrality of the *gaokao*, I found that vocational colleges in Guangdong are nonetheless able to admit students according to their institutional priorities, even when those students do not perform well on the national exam. This flexibility is made possible through a series of reforms initiated by both central and provincial governments in recent years, aimed at "diversifying vocational colleges' student admission approaches."[2] These reforms include the development of a separate *gaokao* track for vocational institutions—one that emphasizes practical skills and technical aptitude rather than the theoretical and academic knowledge prioritized in university admissions.[3]

[2] See "Gaokao gaige nengfou rang gaozhi yinglai 'huangjin shidai'?" (Can the national college entrance examination reforms herald a "golden age" for Vocational Colleges?). *China Youth Daily*, September 29, 2014, https://zqb.cyol.com/html/2014-09/29/nw.D110000zgqnb_20140929_2-11.htm. Accessed July 2, 2025.

[3] See "Guangdong: 2017nian wosheng suoyou gaozhi yuanxiao quanmian shixian fenlei kaoshi zhaosheng" (Guangdong: In 2017, all higher vocational colleges in our province fully implemented categorized admission exams). *Gaokao.com*, November 21, 2016, https://gaokao.chsi.com.cn/gkxx/zc/ss/201611/20161118/1568518145.html. Accessed July 2, 2025.

A particularly significant reform has been the introduction of experimental ***autonomous enrollment*** (*zizhu zhaosheng* 自主招生), which grants selected vocational colleges the authority to design and administer their own admissions processes. In 2017, for example, 50 vocational colleges in Guangdong were allocated a total of 22,609 autonomous enrollment slots, accounting for approximately 20% of their total planned intake. Under this system, each institution develops its own assessment protocol, typically comprising written examinations and interviews, to evaluate candidates on dimensions not captured by the national *gaokao*. Students who pass these institutional evaluations are exempt from taking the *gaokao* altogether. This mechanism enables colleges to recruit students who may lack strong performance in traditional academic subjects—such as mathematics, Chinese literature, and English—but who demonstrate practical aptitude, hands-on learning ability, and motivation aligned with vocational training. As such, autonomous enrollment directly supports the core mission of vocational education by prioritizing skill potential over rote academic achievement.

This policy has been a cornerstone of vocational education reform for over a decade, and its scope has steadily expanded. The increasing quotas allocated to Guangdong's vocational colleges—from 7628 in 2012 to 23,634 in 2016[4]—reflect both growing institutional trust and a deliberate shift toward localized, school-based decision-making in student selection.

While autonomous enrollment enhances access, graduation remains subject to a dual requirement under the "double certificates to graduate" (*shuangzheng biye* 双证毕业) policy: students must earn both a formal school diploma and an NVQS credential relevant to their field of study. The NVQS component introduces a nationally standardized benchmark, which in principle could constrain institutional flexibility. Yet in practice, I found that NVQS certification rarely poses a serious obstacle to graduation. A wide range of vocational schools, training institutes, and even state-authorized testing bodies have developed strategies to ensure high pass rates. Institutions have strong incentives to support student success: a

[4] See "Gaokao kaosheng zhuyi! Jinnia Guangdong gaozhi yuanxiao baokao N ge bianhua ni dou zhidaoma?" (Attention gaokao students! Do you know all the key changes to Guangdong's vocational college applications this year?). *Nanfang Daily*, February 21, 2017, https://static.nfapp.southcn.com/content/201702/21/c292209.html. Accessed July 2, 2025.

low certification or graduation rate could damage their reputation and deter future enrollment. Indeed, during recruitment campaigns, many vocational colleges explicitly promise prospective students that they will graduate with both credentials in hand.

Curriculum Development

Although NVQS certification is generally attainable, allowing it to shape curriculum development would be problematic, as the qualification system is widely perceived as disconnected from actual industry needs (Kong, 2015). My research confirms this critique: across the institutions I studied, the NVQS framework is largely regarded as outdated, irrelevant, or administratively burdensome. As a result, it has been effectively marginalized in the day-to-day training processes of most vocational schools. Teachers and employers alike expressed this view with notable consistency. Rather than structuring entire programs around NVQS standards, schools typically offer short, targeted training modules—lasting a few weeks—immediately before qualification exams to help students pass. These sessions are often treated as supplementary, rather than integral, to the core curriculum.

In contrast to the limited influence of the NVQS, employer input plays a far more decisive role in shaping what is taught in classrooms and workshops—a point elaborated further in Chap. 3. In practice, schools enjoy near-total autonomy in curriculum design. Mrs. Feng, Dean of Academic Affairs at Qingyuan Polytechnic, articulated the prevailing institutional framework:

> In total, programs must consist of between 1400 and 1800 credit hours, with no fewer than 50 percent dedicated to applied or practice-oriented courses, and a weekly cap of 26 classroom hours. As long as departments meet these broad parameters, they have full discretion in designing their curricula. (II: 49)

Despite these formal guidelines, I found that even such minimal constraints are frequently circumvented. At Guangdong Science and Technology Polytechnic, a vice provost explained to a group of collaborating employers concerned about the time devoted to mandatory political and ideological education:

> All courses in the curriculum can be adjusted freely, except for those designated as mandatory by the state. But in fact, we have even restructured many of those. National policy stipulates that political courses must account for 36 credit hours. We now deliver only 18 of those through traditional classroom instruction. The remaining 18 are reclassified as applied courses (meaning students use that time to engage in vocational skill training—noted by the author). This shows that we do have strategies to maneuver within curriculum development. Today, our autonomy in this area is quite considerable. (II: 118)

This exchange reveals that when institutional priorities align with employer demands—particularly the need to maximize time for technical training—schools are capable of creatively reinterpreting or repurposing state-mandated requirements. Such adaptations are not anomalies but reflect a broader pattern of pragmatic institutional agency. Moreover, individual instructors often exercise discretion in textbook selection, with many opting for self-developed teaching materials provided directly by industry partners. These materials are typically more current, context-specific, and technically relevant than standardized textbooks.

In sum, the vocational education system in China today diverges sharply from earlier scholarly characterizations of top-down, state-controlled training. While the central government retains formal oversight—particularly over specialty indexing and certain ideological curricular requirements—local institutions have developed a range of adaptive strategies, some tacitly accepted by authorities, to reclaim autonomy. The state's presence in the daily pedagogical operations of vocational schools has diminished significantly. In its place, schools have cultivated diverse, market-responsive, and employer-engaged educational models. This decentralization has not only enhanced institutional flexibility but also fostered a growing heterogeneity in vocational training across regions and sectors.

Four Skill Development Models in China's Vocational Education

The vocational education system in China exhibits considerable variation across institutions. My research reveals that vocational schools in Guangdong Province coalesce into four distinct skill development models, each with a distinctive combination of four factors: institutional training effectiveness, funding sources, student skill levels, and the scope of labor

Table 2.4 Skill development models of vocational schools

Model	*Funding Sources*	*Training effectiveness*	*Skill level*	*Job market scope*
I. High Performance	High or regular tuition, and state funding	High	High	High-end firms around the country, schools, and research institutes
II. Industry-focused	Regular tuition, state funding, and non-degree training revenues	Medium	Medium-high	Local (provincial) firms, and research institutes
III. Local Market-oriented	Regular tuition, and local state funding	Medium	Medium-low	Local (municipal or provincial) firms
IV. Labor agency	Internship and labor dispatching commission, low or zero tuition	Low	Low	Local or nationwide, but low-end firms

market access. Table 2.4 summarizes the defining characteristics of these four models, and I analyze each in turn.

The High-Performance Model

The first model is what I term the *high-performance model*. Schools in this category achieve strong student outcomes in training effectiveness and skill development, enabling them to target high-end labor markets and secure substantial financial resources—either through high tuition revenues (in the case of private schools) or significant state funding (for public institutions). These schools have historically delivered exceptional results, earning recognition as "national key schools" and receiving extensive state support and social legitimacy. Crucially, they have managed to reproduce this high status over time, reinforcing a self-sustaining cycle of excellence.

Schools under this model equip students with advanced technical competencies, enabling them to secure employment in high-skilled positions at leading firms across the country. Some graduates even enter research institutes or become instructors at other vocational schools. Guangdong Machinery Technician College exemplifies this model. It has a long-standing tradition of training elite students to compete in national and international skill competitions. At the 43rd WorldSkills Competition in 2015, students from this school won five medals, including three of the five golds awarded to the entire Chinese delegation. Four of the five medalists were retained by the school as teachers; the fifth, a champion, was recruited by the China Air-to-Air Missile Research Institute. As the school president explained:

> We have many other students who received the same level of training but did not make it to the final competition. They are also excellent. Many go on to teach at other schools. These students possess systematic knowledge, refined skills, and a strong work ethic and passion. Even when they enter industry, they are placed in core technical roles. (II: 189)

While not every student undergoes this elite training, the achievements of these top performers reflect the overall high training standards of the institution.

Owing to their strong reputation, high-performance schools enjoy a steady influx of applicants, allowing them to maintain financial stability. Baiyun College of Business and Technology is one of China's leading private vocational institutions and has been ranked the top technical school in Guangdong by the Department of Human Resources and Social Security for several consecutive years—surpassing even public counterparts. A striking indicator of Baiyun's success is its tuition rate: while public vocational schools are capped at 3500 yuan per year, Baiyun charges between 11,500 and 14,000 yuan annually. Despite the availability of numerous public alternatives, the school consistently fills its enrollment quotas. A Baiyun administrator proudly attributed this demand to their long-standing reputation:

> It's word of mouth—the result of our relentless focus on student employment and career development. No award compares to the endorsement of our students. (II: 191)

This reputation is reinforced by tangible investments in cutting-edge curricula. A second-year architecture student at Baiyun confirmed this, explaining his decision to enroll despite the high cost:

> This school offers conditions that justify the tuition. I consulted many seniors before enrolling, and they said that as long as you're willing to work, you can land better jobs than graduates from other schools. Recently, the school introduced a course on XXX technology—the most advanced construction estimation software in China. This course is designed to boost our competitiveness in the job market. (II: 197)

High-performance schools offer comprehensive programs, often aligned with high-tech industries and emerging technologies. In just four months in 2013, Baiyun established one of the few 3D printing programs in Guangdong. Given the high cost of industrial 3D printers—ranging from tens of thousands to millions of yuan—the school's investment is particularly notable. In collaboration with two major local printer manufacturers, Baiyun set up two practical training centers equipped with 35 industrial 3D printers and 5 scanning systems. Today, its graduates are employed in R&D, sales, technical support, and management roles across 3D printing firms.

Baiyun's success challenges the assumption that private vocational institutions are inherently low quality. While many low-end, labor-brokering schools are indeed privately owned, Baiyun demonstrates that private ownership can support a high-road strategy in vocational education. Its internship programs are tightly regulated, and exploitative "labor-selling" practices are strictly forbidden. Agreements with partner firms include clear stipulations: interns must be placed in roles relevant to their field of study (no assembly-line labor); overtime is permitted only in exceptional circumstances; and firms must provide specified stipends, career development pathways, insurance, and welfare benefits.

Beyond contractual safeguards, the school's commitment to student welfare is striking. As Department Head Mr. Mo described:

> If a student dislikes their internship position, we change it. If they feel it's not suitable for their development, they can request a transfer. (Author: How often does that happen?) Quite often—some students have asked me to help them switch five or six times. They return, dissatisfied, and we communicate, adjust, and find new placements. This is the 'baby-sitting' work we do—it's the service we offer in exchange for high tuition. (II: 193)

Public schools, constrained by state-mandated tuition caps, cannot generate revenue through fees. However, high-performance public institutions enjoy privileged access to state subsidies and special funding. As Mr. Mo observed:

> State support for public schools is substantial, but it's distributed unevenly. For instance, a special grant for building practical training centers might give one school 100 million yuan, while we receive only 10 to 20 million. That's enough for them to buy 10 to 20 German CNC machines, but only one or two for us. (II: 193)

In return for this preferential treatment, the state expects cooperation in strategic initiatives. In 2000, Volkswagen began establishing its fourth Chinese assembly plant in Foshan, investing 13 billion yuan in its first phase and deploying the most advanced technologies available. By 2013, the factory was expected to operate at full capacity, requiring 4000 skilled workers. Volkswagen turned to the local government for assistance. This high-profile opportunity was offered to four of Guangdong's top public vocational colleges, including Guangdong Industry Polytechnic. As school official Mr. Jie noted:

> We appreciate the opportunity for our students to enhance their skills and improve their incomes. After all, there aren't many companies as good as Volkswagen in China. (II: 8)

My visit to the Foshan plant confirmed the exceptional working conditions: the workshops are highly automated, equipped with advanced robotics and cutting-edge systems—rivaling, and in some aspects surpassing, Volkswagen's facilities in Germany and those of leading Japanese automakers.

Students were sent to Changchun, home of Volkswagen's first joint venture in China, for specialized training. Mr. Jie recalled:

> We sent over 100 third-year volunteers for internships at Volkswagen's headquarters in Changchun. Over six years of collaboration, we supplied more than 2000 students. Changchun functions like a training center—experienced workers mentor newcomers. Then, students return to the Nanhai plant. About 800 were eventually hired. The school invested heavily, mainly to cover the costs of accompanying teachers. Volkswagen did not provide financial support. (II: 8)

According to a Volkswagen manager I interviewed, these hires were placed primarily as welders, with starting salaries between 8000 and 9000 yuan per month—plus generous year-end bonuses. By comparison, Foshan's minimum wage in 2014 was just 1310 yuan per month. This wage premium underscores the exceptional labor market outcomes achieved by graduates of high-performance vocational institutions.

Industry-Focused Model

The second model is what I term the *industry-focused model.* Schools operating under this model possess deep, specialized expertise in a particular industrial sector, rooted in long-standing institutional ties to state administrative bodies and leading state-owned enterprises (SOEs) in that field. These relationships are a legacy of the socialist planned economy, during which many vocational schools were established and directly administered by central ministries responsible for key economic sectors. Although the broader economic system has undergone significant market-oriented reforms, numerous schools have retained—albeit to varying degrees—their historical affiliations with specific industries, SOEs, and the tradition of sector-specific technical education. These enduring connections not only shape the schools' curricula and organizational identity but also enable them to integrate cutting-edge technologies and industry-specific know-how into their training programs.

A prime example is the Guangdong Vocational College of Post and Telecom (GDCPT), a telecommunications-focused institution. Founded in 1949, GDCPT was originally established to train technical personnel and mid-level managers for the post and telecommunications sector under the centrally planned system. From the 1950s to the 1990s, it functioned as a training arm of the Guangdong Post and Telecom Bureau. The sector underwent major structural reforms starting in the late 1990s, including the separation of postal and telecommunications services and, in 2000, the transformation of the Ministry of Posts and Telecommunications into the monopolistic Guangdong Telecom Corporation. Unlike many other formerly state-run institutions, GDCPT remains under the ownership of this corporation, although its academic affairs are now administered by the MOE.

The school currently comprises four departments: Information and Communication Engineering, Mobile Communications, Computer Science, and Economics and Business—each reinforcing its deep alignment with the telecommunications and digital infrastructure sectors. This specialization facilitates continuous knowledge exchange between the institution and its parent industry.

On one hand, GDCPT delivers extensive non-degree, customized training to government agencies, SOEs, multinational corporations (MNCs), and private firms. As a public institution, it cannot raise tuition beyond state-mandated levels, making such industry training a critical revenue stream. In 2008, the college established a dedicated training subsidiary to formalize and expand these services. By 2016, it had been delivering over 300,000 person-days of training annually, generating more than 100 million yuan in annual revenue from this enterprise.

On the other hand, knowledge and technology flow from industry to the school. Since 2004, GDCPT has established six firm-based practical training centers, including partnerships with the Guangzhou Post Bureau and Guangdong Telecom. These leading enterprises have also assigned 80 senior engineers and technicians to serve part-time as guest instructors, ensuring that teaching remains closely aligned with real-world technical demands and operational practices.

Industry-focused schools are most commonly found in highly centralized sectors dominated by a small number of large, often state-controlled firms. In the case of GDCPT, the telecommunications industry remains highly concentrated, with the "Big Three" operators—China Telecom, China Mobile, and China Unicom—controlling the vast majority of the market. Similarly, postal services remain largely state-monopolized. As a result, China Telecom, China Post, and their extensive networks of subsidiaries and suppliers maintain long-term collaborative relationships with GDCPT.

Another illustrative case is Guangdong Communication Polytechnic (GCP), founded in 1956 by the former Guangdong Department of Communications. The school inherited strong institutional ties to both the railway and marine transportation sectors, which continue to be dominated by SOEs. It also developed specialized expertise in transportation management and infrastructure construction. Over time, GCP expanded its industry networks to include both domestic and foreign automakers and their joint ventures. Today, it houses an automobile department with robust partnerships with major firms such as BMW, Bosch, Nissan, and Toyota, enabling students to gain exposure to global production standards and advanced automotive technologies.

Graduates from industry-focused schools are typically equipped with deep, specialized knowledge and high-level technical competencies tailored to their sector. As a result, they enjoy strong labor market access within their respective industries. In 2016, 61.5% of GDCPT's graduates were employed in the telecommunications sector or closely related IT industries. The top ten employers were all telecommunications firms, primarily subsidiaries or suppliers of Guangdong Telecom.

Because industry-focused schools often represent the highest tier of vocational education within their sector, they are capable of producing elite graduates who access high-end job markets. As Professor Guo from GCP's automobile department explained:

> We have students who go on to work in research institutes—this is not typical in other schools' auto departments. For instance, we place students every year at the Guangzhou Automobile Group's Institute of Automobile. (Author: What skills do these employers value?) I think it's their strong manual skills. Even in R&D, there are many practical tasks that require hands-on expertise. (II: 88)

This quote underscores a critical insight: in advanced industrial settings, practical dexterity and applied knowledge are not merely complementary to theoretical understanding—they are often indispensable. Industry-focused schools, by virtue of their embeddedness in real production systems, are uniquely positioned to cultivate this blend of skill and knowledge.

In sum, the industry-focused model reflects a path-dependent institutional strategy, where historical ties to state-led industrial structures have evolved into modern mechanisms for skill reproduction and technological upgrading. These schools function not only as training grounds but also as knowledge intermediaries between state-owned enterprises, regulatory bodies, and the broader labor market. Their success illustrates how legacy institutions can adapt to market reforms without sacrificing their core mission of sector-specific excellence.

Local Market-Oriented Model

The third model is what I term the *local market-oriented model.* Institutions operating under this model exhibit greater flexibility in specialty selection and curriculum development compared to those in the industry-focused model, as they are not bound by historical institutional ties to specific state-owned enterprises or central ministries. However, their training effectiveness and, consequently, the skill levels of their students tend to be lower than those of the high-performance and industry-focused schools, largely due to the absence of deep partnerships with leading firms or access to advanced industrial technologies. Graduates typically enter the labor market at the municipal or district level, rarely securing positions beyond their immediate geographic region.

As a result, these schools adopt a highly pragmatic, market-responsive approach, prioritizing employment outcomes in local labor markets. Their development is significantly shaped by the structure of local product markets—particularly the presence of industry clusters. While this market-driven logic enables schools to adapt quickly to shifting demand, it can also incentivize short-term, cost-minimizing behaviors, such as offering low-investment academic programs or engaging in labor brokerage. Therefore, a key differentiating factor—distinguishing this model from the labor agency model discussed later—is the presence of proactive local state support and strategic intervention, which helps align educational offerings with long-term industrial needs and discourages exploitative or superficial practices.

A central mechanism shaping this model is industrial clustering, which often catalyzes the emergence and consolidation of vocationally oriented institutions. Dongguan offers one of the clearest illustrations of this dynamic in Guangdong Province. As one of the earliest and most industrialized cities in the region, Dongguan's development over recent decades has been characterized by pronounced industrial agglomeration, with major sectors concentrated in its 32 towns and 1 industrial park. Notable clusters include Humen (apparel), Dalang (knitwear), Liaobu (automotive), Houjie and Dalingshan (furniture), Chang'an and Hengli (molds), and Tangxia, Shijie, Shilong, and Chang'an (electronics).

This industrial geography has historically shaped Dongguan's vocational education system. By 2012, 18 vocational schools under the MOE had been established in the city, most located in industrial towns rather than in central urban districts—a pattern distinct from many other Chinese cities where vocational institutions are typically centralized. Table 2.5 lists these schools and their corresponding industry clusters prior to 2012. Of the 18, 11 were situated in major industrial clusters, each funded by the respective town-level education bureau and oriented toward the dominant local industry.

However, in the late 2000s, the Dongguan Bureau of Education (DGBOE) observed a troubling trend: many local vocational schools began introducing so-called fast food specialties—programs in accounting, management, and general business administration. These fields required minimal investment in equipment or practical training infrastructure, unlike traditional manufacturing disciplines, which demand costly machinery and workshop facilities. While the growth of service-oriented programs may reflect broader regional shifts toward

Table 2.5 Dongguan vocational schools and industrial clusters before and after 2012

Before 2012	*After 2012*	*District/Town*	*Industrial Cluster*
School of Business and Economics of Dongguan	School of Business and Economics of Dongguan	Guancheng (central district)	
Dongguan Vocational & Technical School			
Dongguan Sports School	Dongguan Sports School	Dongcheng (central district)	
Dongcheng Vocational & Technical School	Dongguan Commercial School	Dongcheng (central district)	
Nancheng Vocational & Technical School		Nanchen (central district)	
Dongguan Health School	Dongguan Health School	Daojiao	
Dongguan Science & Technology School	Dongguan Science & Technology School	Hengli	Plastic mold manufacturing
Humen Weiyuan Vocational School	Dongguan Textile & Fashion School	Humen	Apparel manufacturing
Dalang Vocational School		Dalang	Knitwear manufacturing
Houjie Technical School	Dongguan Light Industry School	Houjie	Shoes and furniture manufacturing
Chang'an Vocational School	Dongguan Electromechanics Engineering School	Chang'an	Electronics and metal mold manufacturing
Tangxia Science & Technology School	School od Electronics & Technology of Dongguan	Tangxia	Electronic manufacturing
Changping Huangshui Vocational School	Dongguan Electronics & Commerce School	Changping	Logistics and e-commerce
Shijie Vocational School	Dongguan Information Technology School	Shijie	Electronic manufacturing
Shilong Vocational & Technical School		Shilong	Electronic manufacturing
Mayong Vocational School		Mayong	
Liaobu Vocational & Technical School	Dongguan Auto-Technology School	Liaobu	Automobile manufacturing
Dalingshan Vocational & Technical School	Dongguan Furniture School	Dalingshan	Furniture manufacturing

deindustrialization in the Pearl River Delta, DGBOE officials became concerned that schools were adopting these programs primarily for cost-efficiency rather than labor market relevance. As Ms. Zeng, a DGBOE official, explained:

> Schools did this on the one hand to save money. On the other hand, it is appealing to parents who often belittle blue-collar workers. But graduates from these specialties cannot find high-skilled jobs. Accounting students, for instance, mostly end up working as cashiers in supermarkets. Also, development of local industries needs skilled workers. (I: 11)

This observation underscores a tension between market responsiveness and industrial utility. In response, the DGBOE initiated a comprehensive reform in 2012 aimed at restructuring and consolidating the city's vocational education system. The 18 schools were reorganized into 13, with a deliberate effort to strengthen each institution's alignment with its local industry cluster. As Zeng noted:

> We reintegrated and renamed some schools to enhance their local industry focus, and got rid of several "fast food" specialties. The bureau gave them money. We allocate a special funding of 30 million yuan per year for schools' equipment purchasing, but only for their industry-featured specialties, in order to encourage them to reform. (I: 11)

Table 2.5 compares the vocational schools in Dongguan before and after the 2012 reform. The outcome was a more coherent and specialized system: each school now has a sharper industrial focus and deeper integration with its corresponding cluster.

While schools under this model generally lack access to national- or provincial-level funding and prestige, they can still achieve strong financial and institutional support if located in economically vibrant regions, as demonstrated by Dongguan. Conversely, even in less developed areas, determined local state intervention can yield significant results. Qingyuan, one of Guangdong's poorest cities, has a substantial agricultural sector and limited industrial base. Yet the local government has demonstrated exceptional commitment to vocational education. Through generous funding, policy incentives, and administrative support, the Qingyuan Bureau of Education encourages schools to tailor their programs to regional economic needs.

At Qingyuan Polytechnic, officials expressed appreciation for this support and, in turn, have developed specialized programs aligned with local industry, such as an Agricultural Technology and Management program. This reflects a reciprocal relationship between local government and educational institutions—one that prioritizes regional development over generic or speculative curriculum expansion.

Graduates from schools in this model typically attain middle-level technical skills, enabling them to secure stable, moderately skilled employment. At Conghua Technical School, which I classify under this model, I interviewed a recruitment manager from Gree, Inc., one of the world's largest manufacturers of home appliances. Gree recruits students from the school annually. The manager clarified the nature of their placements:

> Our HQ factory has full production chains, but students will not be working on assembly lines. What we need are: First, students in Molding will work in our molding workshops; Second, students in Refrigeration will work on refrigeration detection or assistant refrigeration R&D jobs; Mechanics students will work in our Institute of Electromechanics, the Institute of Automation, or the Automatic Equipment Manufacturing Department—all relevant to their specialties. (III: 1)

This testimony highlights the functional integration between local vocational schools and regional industry. While the skill level may not reach the elite tier of the high-performance model, the alignment with actual production needs ensures meaningful employment and contributes to local industrial upgrading.

In sum, the local market-oriented model reflects a context-sensitive, territorially embedded approach to vocational education. It thrives where local governments actively mediate between education and industry, preventing short-termism and reinforcing long-term developmental goals. Though constrained by geography and resources, these schools play a vital role in sustaining regional-industrial ecosystems and providing viable career pathways for non-elite students.

Labor Agency Model

The fourth and final model is the *labor agency model*. This represents the most problematic end of the vocational education spectrum and aligns closely with the negative stereotypes often portrayed in both academic literature and public discourse. In Guangdong, however, such schools constitute only a small minority, typically limited to underfunded private institutions. These schools operate under what might be described as a "triple scarcity": they lack historically rooted industry connections (a defining feature of the industry-focused model), strong local state support (characteristic of both the high-performance and local market-oriented models), and high social recognition (a key asset of the high-performance institutions).

As a result, they are largely excluded from additional government subsidies, which tend to be directed toward higher-ranked public schools or those demonstrating strategic alignment with regional development goals. Without institutional legacies—whether in the form of elite networks, industry partnerships, or reputational capital—these schools have few resources upon which to build effective training programs. Their educational offerings are often weak, leading to poor skill acquisition among students and limited access to decent employment. This, in turn, undermines student enrollment, forcing many of these institutions to lower tuition below state-stipulated levels to remain competitive.

This downward pressure on revenue creates a self-reinforcing cycle of underfunding and declining quality. To compensate for lost income, some schools resort to treating students as commodities, channeling them into low-wage, labor-intensive jobs at manufacturing firms in exchange for unjustified commissions—a practice commonly referred to as "selling students." This transforms the school into a de facto labor brokerage, entrenching a vicious cycle in which diminished educational investment is offset by commercialized student placement.

In their efforts to attract students, these schools employ a range of recruitment strategies, including offering reduced or even zero tuition. Some go further by allocating a significant portion of tuition revenue to teachers or external agents as performance-based commissions for enrolling students. While these tactics may temporarily boost enrollment, they further strain the school's already precarious finances, increasing the incentive to engage in labor brokerage to cover operational shortfalls.

A well-documented strategy involves sending student interns—often involuntarily—to work on assembly lines in mass production factories. These placements typically occur in sectors facing labor shortages and high turnover, where employers are willing to pay commissions to schools for a steady supply of temporary labor. Despite explicit state regulations prohibiting such practices, some private schools dispatch students for internships as early as their second year of study. This truncates the formal education period, severely undermining the quality and coherence of training. Students placed in such roles, lacking adequate skills, are routinely assigned to low-end operational tasks, reinforcing their marginalization in the labor market.

Both central and Guangdong provincial authorities have recognized the severity of this issue. In Dongguan, for example, the Dongguan Bureau of Education (DGBOE) has implemented two key interventions to curb these practices. First, it has imposed strict limits on internship duration: three to six months for public schools, and no more than one year for private institutions. Crucially, the DGBOE prohibits the placement of second-year students in full-time internships. Second, the bureau has expanded financial support to private schools. While national policy limits tuition subsidies to the first two years of study, the DGBOE provides an additional year of funding, helping to alleviate financial pressure. As one DGBOE official explained:

> The state's support for private schools is not that big. They largely rely on tuition for operation... Many of them used to sell first and second year students to factories as interns, while we have completely forbidden this behavior. Any school that is found to do so will immediately lose their license. (I: 11)

While it remains uncertain whether these measures have entirely eradicated labor-selling practices in Dongguan's private sector, they demonstrate the potential of local state intervention to disrupt exploitative patterns. When combined with regulatory enforcement and targeted financial support, local governments can create conditions that discourage short-term profit-seeking and promote more educationally sound practices.

It is important to note that labor agency schools are more prevalent in China's inland provinces, where economic development is lower and surplus rural labor is abundant. In these regions, many junior secondary school graduates—often from farming households—choose not to pursue academic education but are still below the legal working age of 16. Private vocational schools in these areas frequently serve as intermediaries, recruiting such youth and channeling them into sweatshop-like factories in the industrialized coastal regions. This transregional labor brokerage reflects broader structural inequalities in China's education system, where vocational education in underdeveloped areas functions less as a pathway to skill development and more as a mechanism of labor market insertion.

In sum, the labor agency model reveals the darker side of vocational education when left unregulated and under-resourced. It emerges in the absence of supportive institutional frameworks—where weak state backing, lack of industry integration, and low social legitimacy converge to create conditions conducive to exploitation. Yet, as the Dongguan case illustrates, even schools in this category can be steered toward more constructive roles through deliberate local state strategies that combine financial support with strict regulatory oversight.

What Shapes China's Vocational Schools?

Contrary to what the literature has presented, the four models outlined above—high-performance, industry-focused, local market-oriented, and labor agency—demonstrate significant variation across Chinese vocational schools in terms of training effectiveness, funding sources, and their graduates' skill levels and job market scopes. What accounts for this diversity?

As illustrated, these differences are not random but systematically shaped by four interrelated factors: state support and strategic intervention, local industrial structures, and a school's institutional legacies and ownership type. Table 2.6 synthesizes these explanatory variables across the four models, highlighting how their configuration gives rise to distinct skill development pathways. In this section, I draw on classic literature and the Chinese practices presented above, and theorize how these institutional and organizational factors that a school is endowed with interact to produce the observed patterns. This helps discuss their implications for understanding the political economy of vocational education and skill formation in contemporary China.

State Support and Strategic Intervention

While the state has by and large withdrawn from vocational schools' daily operation and left them sufficient autonomy to devise and carry out their training programs, central and local governments still provide various forms of support to schools and conduct strategic planning that potentially changes the way that schools perform and interact with other players. To begin with, the MOE and the MOHRSS and their all-level branches are the administrator of Chinese vocational schools, and the primary funder of all public ones. Estimated based on the China Statistical Year

Table 2.6 Factors that account for the variations within a taxonomy of Chinese vocational schools

	Institutional Factors		*Organizational Factors*	
	State support and strategic intervention	*Local industrial structures*	*Institutional legacies*	*Ownership type*
High performance	*High national and local supports, and having vocational education lead local industrial development*	Cutting-edge industry influenced	*Social reproduction of high state and social recognition*	Public or private
Industry-focused	Medium-level state support	Single industry focused	*Industry and leading SOE connections*	Most, if not all, public
Local market-oriented	*High local state supports, and localization strategy of the state*	*Local industry cluster embeddedness*	Local reputation and industry connections	Public or private
Labor agency	No or little state support	No industry cluster embeddedness	No industry connections beyond "labor trading"; low social recognition	*Most, if not all, private*

Note: *italics* cells represent dominant factors in the formation of a certain model

Book 2016, approximately 80% vocational schools are public ones, whose graduates accounted for 92.6% of the total vocational school graduates in 2015. Although the 1980s reform of the vocational education system sought to diversify the schools' financial sources by placing more funding responsibilities on the local governments, as well as introducing self and social funding mechanisms, the central government still plays a dominant role and serves as a major provider of school funding (Schnarr et al., 2008). And in the early 2000s, there was a trend of state policies toward recentralizing the financing approach and the supervision of the fund of the educational system in general (Murphy & Johnson, 2009). I found that state support is crucial to sustaining the high-performance model of top-tier public schools. In this case, national and regional governments provide tremendous resources in various forms to schools, in recognition of their achievement, but also in exchange for their cooperation on certain state-initiated projects.

The state provides various supports also to private schools although to a much lesser extent. This is justified by the fact that education is essentially public goods that often entail state direct intervention to remedy a major collective action (free-rider) problem (Streeck, 1989). What I found in Dongguan supports this theory and indicates that state support and regulatory strategies help avoid private schools' short-term behaviors, and potentially move those trapped in the labor agency model up to the local market-oriented model.

The second factor here is local state strategies. The state can enforce various reform agendas to restructure the vocational education system, some of which can engender fundamental changes to the way that practitioners interact with one another. And this role of the state has recently become more significant given the growing importance of vocational education in China's economic reform. In the ongoing apprenticeship reforms documented in Chap. 4, the MOE and the MOHRSS have each adopted a different approach to institution building, but both have imposed a certain level of restructuring on how schools and employers collaborate and train students together. In the abovementioned Dongguan case, the DGBOE successfully enforced a reform over local market-oriented schools to reinforce their integration with local industry clusters. All in all, we expect to see different levels and forms of state support and strategies to be linked to different skill development models of vocational schools.

Local Industrial Structures

Vocational schools are inherently tasked with the mission of serving local industrial development, which entails extra direct input of local employers in their training processes compared to other schools. The dominant industrial relations paradigm provides the insight that product market structures and changes are often reflected in labor market practices (Kochan et al., 1986). As the Chinese vocational education system has become an increasingly important labor market institution for employers to tackle the problem of skilled labor shortage, the linkage between local industrial structures and vocational schools' skill development patterns has been further strengthened. This linkage can be also theoretically derived from the regulation school pioneered by Boyer (1990) and Leborgne and Lipietz (1988), as well as the varieties of capitalism perspective (Hall & Soskice, 2001). Both literatures consent on the fact that skill development systems—in their equilibrium conditions—are often historically proven to be compatible with the dominant production modes of an economy (Green, 1992; Sorge & Streeck, 1988).

Here an often-seen phenomenon is industry cluster embeddedness of vocational schools. In geographers' careful examination of industrial clusters' formation processes, educational institutions are deemed indispensable infrastructures for the emergence and sustainability of a cluster. In turn, these co-located firms, with their economy of scale advantage, provide educational entities with various resources and job market opportunities to incentivize them to customize training programs according to what local employers need (Feldman et al., 2005). In both the United States and Europe, community colleges are found to intentionally target industry clusters in which they are embedded. An industry cluster benefits from vocational schools continuous supply of skilled labor, but also actively shapes the local vocational education system through transferring technology and know-how, as well as constructing formal and informal learning contexts for trainees (Rosenfeld, 2000).

But even without salient industry clustering, both community colleges in the United States (Dougherty & Bakia, 1999) and vocational schools in China (Chap. 3) are found to have strong local industry orientation, actively collaborating with important firms to serve dominant regional industry needs. It is therefore no surprising to see that local industrial structures, especially dominant industries and leading firms, play important roles in the formation of Chinese vocational schools' skill development models. In this chapter, I found in Dongguan that local market-oriented vocational schools have historically emerged and embedded in district industry clusters, and therefore each have a strong clustered-industry focus.

Institutional Legacies

Institutional legacies are generally important for explaining organizational outcomes. This perspective is supported by resource-dependence theory, which views institutional legacies as critical external resources that organizations can leverage to shape their strategies (Pfeffer & Salancik, 1978). In the context of Chinese vocational schools, two institutional legacies of a school have turned out to be especially significant.

The first is reputation. Prestigious schools are able to secure a stable supply of students and attract higher-achieving applicants (Monks &

Ehrenberg, 1999). Moreover, reputation can be socially reproduced: it functions as a valuable organizational attribute that contributes to high performance, which in turn reinforces social recognition and sustains rapid institutional growth. This dynamic has been extensively studied in the context of educational organizations, particularly American business schools, which are frequently cited as illustrative examples in management literature (Boyd et al., 2010; Corley & Gioia, 2000; Pfeffer & Fong, 2002). Scholars have also observed in the United States higher education system that reputation enables top-tier institutions to charge premium tuition (Peters, 2007)—a pattern consistent with my findings regarding certain private "high-performance" vocational schools in Guangdong.

Reputation is often institutionalized through school rankings, which in China are closely tied to state recognition and generous public support. In higher education, state funding has been shown to disproportionately favor elite universities such as Tsinghua and Peking University, in pursuit of international prestige and competitiveness (Ngok & Guo, 2008; Yang & Welch, 2012). I find that this same logic extends to the vocational education sector.

The second key institutional legacy, particularly distinctive to many Chinese vocational schools, is their historical industry connections rooted in the socialist centrally planned economy. Since the founding of the People's Republic, vocational schools were established and operated by sector-specific ministries (e.g., the Ministry of Textile Industry) and/or their state-owned enterprises (SOEs), with the mission of training cadres and technicians for national industrialization. By 1992, 4392 technical schools had been established—approximately half by enterprises—producing 5.3 million technicians, most of whom were employed in SOEs (Xu, 1993). The central government's bureaucratic reforms during the 1980s and 1990s first abolished these sector-specific ministries,[5] transferring ownership of their affiliated schools to state-owned enterprises (SOEs) that evolved from the former ministries. Later, in the late 1990s, SOEs were required to reduce their social functions and divest their affiliated schools in a few policies by the central government. The result was that

[5] See "Naxie xiaoshi de buwei" (The disappeared ministries). *People Digest*, 2013(4), https://paper.people.com.cn/rmwz/html/2013-05/01/content_1247974.htm. Accessed July 2, 2025.

ownership of most vocational schools had been largely transferred to the MOHRSS or the MOE by the 2000s.[6]

In practice, although most public vocational schools in China are now formally administered by the Ministry of Education (MOE) and the Ministry of Human Resources and Social Security (MOHRSS), many institutions continue to maintain varying degrees of connection with their founding state-owned enterprises (SOEs), several of which remain industry-leading firms. This pattern is most evident in highly centralized sectors characterized by SOE dominance or oligopolistic structures, such as telecommunications, construction, railway, and marine transportation. Consequently, numerous Chinese vocational schools operate within an industry-focused model, where their development and financial sustainability are significantly dependent on leading firms, their subsidiaries, and supply chains.

Ownership

Ownership itself does not directly determine a specific skill development model, but it is closely associated with the other three institutional and organizational factors discussed above. Privately owned vocational schools often face what I term "triple scarcity"—a lack of state support, high state and social recognition, and historically rooted major industry connections or deep embeddedness in local industry clusters. As a result, some private schools resort to labor-brokering activities to offset the high costs of student recruitment.

In the Chinese context, private vocational schools typically receive significantly less state support and lack the inherited SOE connections common among public institutions. They also tend to enjoy lower social recognition compared to their public counterparts. Furthermore, alumni endowment—a key revenue source for American private institutions—is not a well-established tradition in China, and where such endowments exist, they are concentrated in public universities and academic colleges. Therefore, private ownership of vocational schools is strongly associated with the labor agency model.

As analyzed above and summarized in Tables 2.4 and 2.6, the skill development approaches across the 25 schools exhibit distinct patterns.

[6] A comprehensive review of this reform history and relevant policy documents is available at: https://zhuanlan.zhihu.com/p/455947824. Accessed July 2, 2025.

While these patterns emerge from unique historical, economic, and political conditions, their fundamental contours reflect the interplay of four key institutional and organizational factors, with the relative influence of these factors varying across schools. Table 2.6 illustrates how the four skill development patterns arise from different combinations of these factors, highlighting the dominant factors for each model.

Specifically, ***high-performance schools*** are characterized by high training effectiveness, producing graduates equipped with relatively advanced skills. These graduates typically secure positions in high-end firms as technicians or, after several years, as managers. Some also pursue teaching or research roles in other vocational schools or research institutes nationwide. In return, high-performance private schools can charge tuition above state-mandated levels while still maintaining strong enrollment. Public high-performance schools, while unable to raise tuition freely, receive substantial state financial support that enables their advanced vocational training model.

The dominant factors for this model are a school's high social recognition and substantial state support, which are often interlinked with its reputation. This reputation is self-reinforcing and helps sustain long-term performance, while also creating a significant barrier for other schools attempting to catch up. This dynamic aligns with the concept of latecomer disadvantage in strategic management literature (Lieberman & Montgomery, 1988; Barney, 1991), where "the rich get richer, and the poor get poorer" (Corley & Gioia, 2000). Because these schools target national and provincial labor markets, they are less influenced by local industrial structures and thus tend to offer comprehensive curricula. Nevertheless, cutting-edge industries shape their development, as these schools frequently introduce up-to-date programs aligned with the latest industrial trends to maintain their elite status.

In comparison, the ***industry-focused model*** exhibits a different logic. These schools concentrate their curricula on a specific industry or closely related sectors. Their training effectiveness can be exceptionally high—sometimes representing the highest skill levels within that particular sector—due to their historical ties with leading firms. Graduates are typically recruited directly by these leading firms, their subsidiaries, and suppliers. Most, if not all, of these schools are public institutions and therefore cannot charge tuition above state-set limits. However, a crucial revenue stream comes from non-degree training programs offered to industry partners.

These characteristics stem from a significant institutional legacy: many of these schools were originally established by sector-specific ministries during the socialist centrally planned economy. Although most formally detached from their founding enterprises by the early twenty-first century, they have retained, to varying degrees, historical connections with leading SOEs. As a result, they have inherited valuable resources in employment networks, technological expertise, and institutional roles as external training units for these firms.

Local market-oriented schools represent middle-range or lower-tier institutions. Their key distinction from high-performance schools is the absence of widespread recognition. Consequently, they primarily serve regional labor markets—within cities, districts, or occasionally provinces—and focus intensively on local employment outcomes. Rather than targeting cutting-edge industries or cultivating specialized talent, they aim to produce a middle-range workforce that is immediately useful to local firms. This orientation makes them highly adaptable in terms of specialties and curricula, demonstrating a significantly stronger local market focus than the two previous models.

Local industrial structures are often mirrored in these schools' program offerings and curriculum design. Historically, many emerged within and became deeply integrated into regional industry clusters, developing expertise grounded in those local industries. Another critical factor is local state support and industrial strategy. As illustrated in the Dongguan case, the city's education bureau used both administrative authority and financial incentives to strengthen vocational schools' alignment with local industry clusters. Even in cities without strong clustering—such as Qingyuan—local government support often encourages schools to meet the needs of dominant regional industries. Moreover, local state backing helps prevent short-term opportunistic behaviors. While all middle- and low-tier schools possess an inherent tendency toward local market responsiveness, it is local state intervention that often prevents them from sliding into the labor agency model and engaging in unjustified labor brokerage practices.

Finally, *labor agency schools* are defined by both low-quality training and active involvement in labor-selling activities. These institutions occupy the lower tier in terms of training effectiveness and student skill development. They are classified as such not only because of poor educational outcomes but also because they operate businesses that sell student interns and graduates to low-end firms as cheap labor, often in exchange for illegitimate commissions. Many send students to work during their second year of study, further compressing an already ineffective education process. As a result, their primary labor market function becomes labor dispatching, not education. It is more accurate to describe them as quasi-labor agencies, and their output as cheap labor rather than skilled workers. While some such schools exist in Guangdong, they are most commonly found in inland China, where inter-provincial labor brokerage is widespread. Indeed, Smith and Chan et al. (2015) and Su (2010–2011), in their Foxconn-centered research, identified student interns primarily from inland provinces such as Henan, Anhui, Hubei, Sichuan, and Guangxi.

These schools are typically poorly funded private institutions trapped in a "triple scarcity": lacking support from national or local governments, lacking deep-rooted connections with key industries or regional clusters, and lacking sustainable social recognition. They rely heavily on tuition revenue, but due to low reputation, they incur high recruitment costs—often resorting to reduced or even zero-tuition policies to attract students. To compensate for these losses, they turn to labor-selling activities.

It should be noted that private ownership is neither necessary nor sufficient for the labor agency model. Baiyun, for example, demonstrates how private schools can leverage market mechanisms to achieve high performance. Moreover, in less regulated regions of inland China, some public schools may also engage in labor-selling practices. However, I find that most labor agency schools in Guangdong are privately owned. This is likely because private ownership is frequently associated with the aforementioned triple scarcity.

The literature has identified several serious issues in certain Chinese vocational schools, including low training effectiveness and unethical intern-placement practices. While most schools I studied in Guangdong do not conform to the labor agency pattern, I agree that there is room for improvement in training quality and that internship programs—varying by institution—require stronger regulation. Even in the widely admired German apprenticeship system, employers are sometimes found to abuse the firm-based training component by assigning apprentices tasks "non-related to stipulated training agendas."[7] There is ongoing concern about training quality, and German stakeholders continue to seek improvements

[7] See the DGB (German Confederation of Trade Unions) 2016 Union Report, Ch.3 and following, available at: https://www.uebergangschuleberuf.de/wp-content/uploads/2022/01/dgb_jugend_ausbildungsreport_2016.pdf. Accessed July 2, 2025.

through state and social governance mechanisms. Yet these challenges have not prevented the German system from becoming a globally recognized model of national skill development.

Indeed, following widespread media coverage and online dissemination of such issues, the MOE and the MOHRSS jointly implemented a series of regulatory policies in 2016 to govern the use of student interns. While the long-term impact of these measures remains to be assessed, a more comprehensive understanding of the Chinese vocational education system is essential. Such a perspective is indispensable for identifying feasible pathways to upgrade labor agency schools through enhanced state support and strategic direction, and greater involvement of leading firms.

Summary

In this chapter, I have developed an explanatory framework that focuses on the underlying institutional and organizational factors shaping skill development patterns across different Chinese vocational schools. I argue that, in the Chinese context, skill development models emerge from a combination of two institutional factors—state support and strategic intervention, and local industrial structures—and two organizational factors—a school's institutional legacies and its ownership.

Case studies of 25 schools in Guangdong demonstrate that the relative importance of these endowments varies significantly across institutions. Typically, it is one factor, or the interplay of two, that acts as a dominant force in determining a particular skill development pattern. These cases further reveal that different configurations of these factors lead to distinct trajectories in skill formation. Based on this analysis, I identify a taxonomy of four distinctive models of Chinese vocational schools: *high performance*, *industry-focused*, *local market-oriented*, and *labor agency*.

This framework addresses a significant research gap in the scholarship on Chinese vocational education. The decentralization reform and prevailing but inadequate stereotype of vocational schools as mere labor agencies has implicitly pointed to variation across institutions—an aspect previously unexplored in academic literature—and thus opened the critical question: What factors explain this variation? My findings show that substantial differences exist in Guangdong regarding schools' training effectiveness, funding sources, students' skill levels, and job market scopes. While some vocational schools do conform to the labor agency pattern—consistent with existing stereotypes—the majority do not. Instead, most schools have adopted relatively advanced, albeit diverse, skill development models that go well beyond the labor agency archetype.

This research aims to challenge and move beyond the reductive stereotype of Chinese vocational education. However, it is important to clarify that by identifying four skill development models, I do not intend to impose a new set of stereotypes. The taxonomy reflects a spectrum of schools observed in Guangdong, but given the vast regional disparities in China's industrial development and local state strategies, the vocational education system is likely to exhibit significant local variation. The Guangdong experience enables me to demonstrate the existence of such variation and to offer a preliminary explanatory framework. Nevertheless, the nuances identified here require further empirical validation in other regions of China.

Future research should seek to identify and compare potentially different skill development patterns across other parts of the country, as well as the unique local factors that shape them. Indeed, both Chinese (Cooke, 2005; Durden & Yang, 2006; Lai & Lo, 2006) and Western literature (see Ashton et al., 2000 for a review) have largely focused on national-level analyses of skill development systems. While these studies provide valuable insights—particularly from an international comparative perspective—I encourage institutional scholars to pay greater attention to subnational variation. Capturing and explaining such variation is essential for a more complete and accurate understanding of how skills are produced in decentralized, multi-layered educational and economic environments.

CHAPTER 3

Collaboration: How Firms Facilitate the Transformation of Vocational Education in China

The previous chapter demonstrated the central role of decentralization in vocational education reform since the 1980s, which entailed the withdrawal of state authority from the daily operations of schools and resulted in a diversity of development approaches to vocational school development in China. While this reform created important institutional conditions for firms to experiment with innovative training models, decentralization alone does not guarantee that training processes are better aligned with industrial needs. This chapter examines the other side of the decentralization story: the state's retreat occurred alongside another significant trend in China's economic reform since the 1990s—the emergence of a secondary labor market. This development has prompted firms to step in and proactively seek collaboration with vocational schools to secure the skills they require.

Labor markets in China have undergone gradual but steady transformation. Prior to the 1980s, employment systems in state-owned enterprises—commonly known as the "iron rice bowl"—were characterized by an internal labor market model for training. Firms developed firm-specific skills through extensive internal training and retained these workers through lifelong employment arrangements (Walder, 1988). As China transitioned toward a socialist market economy, its growth became heavily reliant on low-cost, labor-intensive, export-oriented manufacturing and the employment of a large migrant workforce of unskilled labor. In

H. Zhang, *Upskilling the Chinese workforce?*,
https://doi.org/10.1007/978-981-95-2656-7_3

response, new labor market institutions emerged, such as labor dispatch agencies, which supplied low-skilled workers for both core and peripheral positions (Xu, 2008; Feng, 2019). These intermediaries enabled firms to access labor while avoiding the job security obligations and legal costs associated with internal training (Liu, 2015).

Vocational schools constitute another significant type of labor market institution. Some scholars view them as de facto labor agencies that dispatch low-skilled, seasonal labor—graduates or student interns—to firms in exchange for questionable commissions (Smith & Chan, 2015; Chan et al., 2015). However, as Chinese firms have sought to move up the value chain and avoid the "middle-income trap" in recent years, they have faced a growing shortage of adequately skilled labor. At the same time, the decentralization reform has granted vocational schools greater freedom to experiment with and articulate new training models. I find that many schools have transformed from mere labor intermediaries into providers of more advanced, firm-specific training. While the previous chapter has revealed three relatively advanced development models adopted by vocational schools beyond labor brokerage, successfully transitioning to or across these models entails significant structural changes at the institutional level. In contrast, the active collaboration with firms presented in this chapter highlights a more practical and micro-level transformation process—one that is widespread among vocational schools in contemporary China.

This transformation has been, to some extent, driven by Chinese firms' increasingly strategic approach to human resource management. Firms have recognized that "making" skills internally compromises their "numerical" flexibility (Lepak & Snell, 1999), while "buying" skills from vocational schools or labor dispatch agencies fails to meet their growing demand for firm-specific competencies. Consequently, I argue that Chinese firms, through collaborative partnerships, have reshaped existing vocational education institutions to externalize firm-specific skills training into workers' pre-employment skill formation. This allows firms to secure a sufficient supply of skilled labor while maintaining employment flexibility. The model illustrates how internal labor market functions can be transferred externally and carried out by organizational partners. How does this process operate, and what are the key mechanisms facilitating the transformation of vocational schools?

To address this research question, I examine a prominent case of transformation in China's vocational education system: the shift from supplying general, low-skilled labor to a broad range of firms toward dedicated partnerships with specific companies—namely, targeted and customized training. My field research focused on three industries: auto maintenance, robotics manufacturing, and beauty services. Each represents a significant sector in contemporary China—traditional manufacturing, modern manufacturing, and modern services, respectively. The latter two have been actively promoted by the state in recent years as part of industrial upgrading and restructuring strategies. While Guangdong has experienced a general deindustrialization trend in its traditional sectors, the auto industry has remained an exception, consistently regarded as a pillar industry in the regional economy. Thus, while I do not claim that customized and targeted training is a universal practice across all regional industries, these three sectors serve as diverse cases (Patton, 1990) that allow for comparison and the identification of common themes, thereby enhancing the generalizability of this increasingly significant model of vocational education.

More importantly, these industries have commonly faced secondary labor market failures—specifically, an insufficient supply of skilled labor in Guangdong. A structural shortage of skilled workers has affected Guangdong and other parts of China since the turn of the century. According to an official report, the Guangdong labor market experienced an average monthly shortage of 175,000 skilled workers during the first three quarters of 2015.[1] Moreover, firms increasingly struggle to recruit workers with the necessary skills from the external labor market. This shared demand for skilled labor has prompted employers, especially leading firms, to seek collaboration with vocational schools. Firms in the automobile, robotics, and beauty sectors, unable to source skilled workers through local labor markets or other intermediaries (e.g., temporary agencies), have adopted collaborative programs with vocational schools as a strategic response.

Specifically, individual schools and firms co-establish school-based training programs focused on preparing students for a particular employer. These programs are commonly known as "customized classes" (*dingdanban*, 订单班) or "sponsored classes" (*guanmingban*, 冠名班). With the

[1] See "Dongguan gongren fangqi 8 k yuexin gongzuo, zhuanxing jiqiren jigong" (Dongguan workers give up 8 k RMB monthly salary jobs to transition into robot technicians). *twenty-first Century Business Herald*, June 8, 2016, https://tech.china.com/news/prnasia/11082256/20160608/22831824_all.html. Accessed July 2, 2025.

exception of state-mandated courses and a limited number of theoretical subjects, students in these programs have little to no exposure to technologies and equipment from other brands throughout their training. The considerable autonomy afforded by the state's ongoing decentralization of the vocational education system enables this distinctive collaborative model to function with minimal institutional constraints. Typically, under the "training for a targeted brand" model, the individual collaborating firm—not the state or any industry-level governing body—dominates the training process within the vocational program.

In the following section, I delineate the motivations and mechanisms underlying collaborative training in these three industries. I identify three variants of the model: "training for a targeted brand" in auto maintenance, "training for a targeted value chain" in robotics, and "training for a targeted market" in beauty. Each variant is embedded within a distinct sectoral labor and product market context, leading firms to exhibit different motivations for collaboration. Despite these variations in incentives, three common themes emerged from my qualitative analysis, constituting the core mechanisms of these collaborative programs: curriculum customization, knowledge transfer, and technology transfer. These findings are summarized in Table 3.1. I argue that, when effectively implemented, these elements have the potential to move China's vocational education system beyond the low-skill "labor dispatching" trap highlighted by scholars (Smith & Chan, 2015; Chan et al., 2015), thereby supporting the broader industrial upgrading agenda of China's economic reform.

Auto Maintenance: Training for a Targeted Brand

Internal Needs for Brand-Specific Skills

The primary motivation for branded automakers to engage in collaborative training partnerships with vocational schools is the demand for brand-specific skills within their own and franchised vehicle maintenance networks. In the automobile industry, firm-specific competencies constitute a significant portion of the essential skillset required of technicians. The ratio of specific to general skills varies across sectors; for instance, it is considerably lower in electronics assembly, where operational workers perform standardized tasks across different companies—such as Foxconn and Flextronics—under the Taylorist production model. In contrast, within the auto industry, vehicle models differ substantially across brands. In

Table 3.1 Three key mechanisms that facilitate the targeted training model

		Auto maintenance Training for a targeted brand	*Robotics/3D printing Training for a targeted value chain*	*Beauty Training for a targeted market*
Key motivations		Internal needs for brand-specific skills	Internal and external (robot buyers') needs for brand-specific skills	Needs for market expansion for their own products
Mechanisms	Curriculum domination *Insertion of firm-specific modules*	Medium	Medium	High
	Knowledge transfer *Textbook and training materials transfer, teacher training and dispatch*	Medium	High	High
	Technology transfer *Training center co-building and equipment transfer*	High	High	High

relatively advanced manufacturing environments, multiple car models are often assembled simultaneously, requiring workers to be familiar with the components of various models they handle. A technician in a 4S (sales, spare parts, service, and survey) store, however, must possess even deeper expertise, mastering the full range of models within a single brand to effectively diagnose, recommend, and execute repairs and maintenance.

Consequently, while a general shortage of skilled labor affects employers across sectors, automakers face an even more acute constraint: qualified maintenance technicians with brand-specific know-how are particularly scarce due to the high proportion of firm-specific skills embedded in their required skillset. Moreover, branded automakers typically invest substantial time and resources in firm-based training to develop a competent maintenance workforce. This often involves establishing dedicated training centers that serve not only their corporate and franchised 4S stores but also authorized third-party repair facilities. In addition, 4S stores themselves bear significant on-the-job training responsibilities, frequently expressing frustration over the time and effort required to bring new hires up to full operational independence. In an interview, a manager at an auto maintenance firm estimated that, without collaboration with vocational schools, it takes approximately three months for a new technician to perform tasks under the supervision of senior staff. By contrast, a workshop supervisor from Volkswagen noted that preparing a new assembly-line worker to operate independently requires only one week of initial training, followed by an additional one to two weeks of on-the-job mentoring. As a result, qualified auto maintenance technicians are among the most sought-after workers in Guangdong's labor market, driven by the region's continuously expanding consumer base.

Mechanisms

In response to these challenges, automakers have adopted a strategic approach: shifting a portion of their traditionally in-house training to vocational schools. This strategy encompasses co-developing curricula, establishing on-campus training facilities, and transferring technologies, training materials, and vehicle models to partner institutions. In turn, schools are highly receptive to customizing their programs to meet the needs of collaborating firms.

The first mechanism is curriculum customization. Schools actively collaborate with partner firms to co-develop specialized programs and course content. At the outset of a partnership, schools consult with the firm to determine which curricular components should be included. They are generally willing to integrate modules preferred by the firm into their academic offerings. As a senior manager from Volkswagen (Foshan) explained:

> Maintenance training requires more knowledge and skills than assembly-line work, and our company has a defined set of standards and competencies. Schools previously faced a challenge—they didn't know what we needed because they were unaware of where students would ultimately be employed. That's why we collaborate on training. Many schools only teach general skills: basic automobile knowledge, vehicle structures, assembly principles, and mechanical drawing. In our collaborative programs, we incorporate brand-specific content—such as the electrical systems of Audi vehicles and Audi-specific maintenance procedures. (III: 69)

He estimated that through collaboration, approximately 20% of the firm-specific training required for a maintenance technician is now delivered within the school's curriculum.

This practice is widespread in the auto industry. A single vocational school may partner with multiple automakers, offering distinct programs tailored to each brand. For example, at the School of Automotive Application at Guangdong Communication Polytechnic, students in their first year take mandatory general education courses—such as political and ideological studies, mathematics, Chinese literature, and computer science—alongside foundational automobile theory. Upon entering their second year, they select a brand-specific program, such as BMW, Toyota, Nissan, Bosch, or Volvo, where they begin intensive training focused on the technologies and models of that particular brand. Throughout this process, students have little to no exposure to the technologies of competing brands.

The second mechanism is technology transfer from firms to schools. Collaborating companies frequently provide schools with critical equipment used in their operations. A common strategy is the co-establishment of on-campus brand-specific training centers. These centers are equipped with key tools, vehicles, and machinery—such as cars, diagnostic systems, or 3D printers—supplied by the partner firm. Most, if not all, applied coursework is conducted within these facilities. At Guangdong Communication Polytechnic's automobile school, for instance, I observed at least five such centers—dedicated to BMW, Bosch, Nissan, Toyota, and Volvo—each stocked with the latest vehicle models and equipment from the respective partners. Students select their brand specialization at the start of their second year and are grouped into brand-specific classes. In core specialty courses, they exclusively use vehicles and systems from the collaborating brand, receiving little or no training on competing brands.

The third mechanism is knowledge transfer, which complements technology transfer. This often involves firms providing systematic training for school instructors. At Guangdong Communication Polytechnic, the head of the automobile department described how BMW trained their faculty:

> BMW is a prime example: our teachers have undergone training and are now certified as BMW trainers—on par with their own internal trainers. Several of our instructors have obtained the highest-level trainer certifications, making them highly skilled. So, they are both schoolteachers and BMW-certified trainers. They occasionally teach at BMW's own training centers. They

> renew their BMW certifications annually. We have a BMW training center on campus, so BMW also sends its employees here for training. (II: 88)

These mechanisms—curriculum customization, technology transfer, and knowledge transfer—are frequently employed simultaneously to ensure the effective integration of firm-specific skills into the vocational education process.

Outcomes

As a result of this collaborative training model, graduates from these targeted programs enjoy significantly better employment prospects in their field of study compared to their peers in non-specialized programs, many of whom end up in jobs unrelated to their training. A program director estimated that the placement rate in specialty-related positions for students in these collaborative programs exceeds 70%, substantially higher than the school-wide average, which typically falls below 50%.

Robotics Manufacturing: Training for a Targeted Value Chain

Internal and External Needs for Brand-Specific Skills

Robotics stands out as an emerging industry in Guangdong, where I observe a growing demand from employers for collaborative training programs. In recent years, local governments have introduced a series of public policies and subsidies aimed at promoting an industrial upgrade agenda focused on "replacing workers with robots." While robotic arms are the most visible form of industrial automation, the broader definition of robotics also encompasses advanced technologies such as 3D printers, AI-powered inspection systems, and autonomous logistics robots. By 2015, Guangdong had already gathered 159 major robot manufacturers. More importantly, both local and external robot firms supply robotic arms to a diverse range of manufacturing sectors—ranging from electronics and automobile to textiles and appliances—thereby replacing thousands of manual laborers. According to provincial government estimates, approximately 23,000 robots were deployed in Guangdong workplaces by 2014, displacing around 151,000 operational workers. By 2015, these figures had risen to 41,000 robots and 269,000 displaced workers. The government projected that robot deployment would reach 300,000 units by 2020, potentially displacing up to 1.95 million workers (Xue, 2016).

The exponential integration of robotics into traditional manufacturing necessitates a fundamental shift in workforce skill composition. On average, the adoption of robotic systems in conventional production environments reduces operational jobs by 19%, while simultaneously increasing the demand for skilled technicians by 17% (Xue, 2016). Employers now require personnel capable of installing, programming, troubleshooting, and maintaining robotic systems—tasks demanding technical expertise far beyond the capabilities of traditional factory workers.

This growing demand for skilled technicians is exacerbated by a critical failure in the external labor market. In more mature industries such as automobile manufacturing, firms can often address talent shortages through labor poaching—hiring experienced technicians from competitors and providing targeted retraining. However, this strategy is largely impractical in the robotics sector, which remains nascent and lacks a sufficient pool of qualified professionals in the regional labor market. Retraining a conventional technician to become proficient in robot maintenance requires significantly more investment than transitioning a technician from one automobile brand to another, such as from Toyota to Volkswagen. The know-how required is not only more complex but also highly brand- and system-specific. This skills gap has become a significant bottleneck, limiting firms' ability to scale up automation. As observed by a local labor department official:

> Replacing operational workers with robots is an inexorable trend of industrialization. But robots also need workers to maintain and repair them. The demand for skilled labor in maintenance is enormous—not only in robot-using firms but across the broader manufacturing sector. These roles come with a high skill threshold. As robots become increasingly prevalent, we are facing a serious labor shortage. (I: 8)

This workforce shortage poses a direct threat to the region's industrial upgrading ambitions, raising concerns not only for policymakers but also for robotics manufacturers themselves. In response, many firms are proactively engaging with vocational institutions through collaborative training initiatives. They are adopting what can be described as a "training for a targeted value chain" model—a strategic evolution of the more narrowly defined "training for a targeted brand" approach. Under this model, robot manufacturers have strong incentives to train technicians not only for their own internal needs but also for the broader ecosystem they lead, including current and prospective clients who purchase and operate their equipment. As a result, Guangdong has seen a surge in robot-related training programs, often co-developed and co-delivered by industry leaders and educational institutions. A similar dynamic is evident in the 3D printing sector, where leading equipment suppliers partner with schools to cultivate a talent pipeline aligned with both their own service needs and those of their client base.

The Mechatronics Department at Guangzhou Baiyun Institute of Business and Technology exemplifies this trend, offering specialized programs in both robotics and 3D printing in collaboration with branded equipment providers. The department chair articulated the multifaceted motivations driving these partnerships:

> A key incentive for Jiehe Company is that we train students who effectively serve as ambassadors for their printers. Because our students are trained on Jiehe machines, we become a steady supplier of skilled labor and a de facto training hub for their ecosystem. Winbo Company operates similarly—our graduates form a major recruitment pool for them each year. I can provide as many as they need. Then there's Hurco, an American firm that entered the Chinese market in 2012 with their advanced five-axis robotic systems. After negotiations, we became their strategic education partner in South China. Many companies purchased Hurco machines but lacked the expertise to operate them. Programming and operation require rigorous, specialized training. Without skilled users, the investment in high-end equipment goes to waste. To address this, we reached an agreement: Hurco donated equipment to us at no cost, and in return, we deliver a comprehensive talent supply solution. (II: 193)

Mechanisms

These collaborations typically begin with joint curriculum development. Mr. Zhang, who oversees a department with partnerships in both automobile and robotics manufacturing, contrasted the nature of these engagements:

> Nissan contributed around 500–600 credit hours of coursework and two proprietary textbooks to our curriculum, specifically designed to train

> maintenance technicians for their production lines. Initially, they sent their own instructors to teach these courses. Over time, our faculty learned the material and now deliver it independently. Their training emphasizes speed, safety, and procedural compliance—how to repair machines efficiently, ensure operational safety, and complete required documentation. A robotics manufacturer, by contrast, tends to focus on highly specific training in system design, installation, and debugging tailored to their own equipment. (II: 174)

The second key mechanism is technology transfer and infrastructure support. At Baiyun Institute, I visited two branded 3D printing practical training centers. One was co-established with Jiehe 3D Printer, which provided 20 printers, accompanying software, and five 3D scanning systems free of charge. The other was developed in partnership with Winbo, which donated 15 printers and associated programs. These in-kind contributions enable schools to offer hands-on training with industry-standard equipment, bridging the gap between theoretical learning and real-world application.

Third, employer-provided instructional materials play a crucial role, especially in emerging fields where standardized textbooks are unavailable. As Mr. Zhang explained:

> Since this specialty is new, every school is essentially developing its own teaching materials. There are no widely accepted textbooks yet, so we either write our own or use materials provided by the firms. The challenge is that no single textbook fits all systems. Our robotics curriculum is built around specific equipment models, making the content highly compatible with that brand—but not transferable to others. If you try to apply these materials to a different brand's system, they simply won't work. (II: 174)

In some cases, firms also dispatch their engineers to teach directly in vocational programs. In the robotics and 3D printing programs at Baiyun, company engineers serve as part-time instructors. For example, Mr. Kong, a senior engineer at Jiehe, is sent annually to deliver a one-week intensive course covering installation, debugging, maintenance, and user best practices for Jiehe's 3D printers. This direct knowledge transfer ensures that training remains closely aligned with current industry standards and technological updates.

Outcomes

While high-level R&D positions in robotics typically require postgraduate degrees (Ph.D. or master's), vocational school graduates are increasingly finding roles in the mid-tier segment of the industry. These include positions such as robot installers, operators, maintenance technicians, programmers, and technical sales representatives. Notably, some vocational graduates are also being assigned tasks traditionally associated with R&D support, such as mechanical design, 3D modeling, and even material development—highlighting the expanding scope and complexity of technical roles accessible through vocational pathways.

Through these collaborative arrangements, firms achieve more than just curriculum influence—they gain preferential access to talent. Although companies are not obligated to hire the students they help train, they are typically granted first-choice recruitment rights. This allows them to select top-performing graduates before other employers, ensuring a steady influx of job-ready, brand-specific talent while retaining numerical flexibility. Remaining students are then placed by the school with other industry partners. Students, in turn, are often inclined to join collaborating firms as their first employment choice, both because these firms are typically market leaders and because much of their training is directly applicable only to the equipment and systems used by those partners. General Manager Li of Jiehe shared insights into their 2016 recruitment of five prospective graduates:

> They will be assigned to two main roles. First, mechanical R&D support—assisting our engineers in product development. Second, technical service in our newly established 3D printing service center for healthcare. In this role, they will work directly with hospital clients. For example, when we receive a digital model from a hospital, these technicians will develop a processing solution, program the print job, and oversee the entire production process to deliver a customized medical device. (III: 75)

This integrated model—where training, technology, and recruitment converge around a specific brand and its ecosystem—demonstrates a strategic alignment between education and industry in Guangdong's emerging robotics sector. It reflects not merely a response to labor shortages, but a proactive reconfiguration of the skill formation system to support technological innovation, industrial competitiveness, and sustainable economic transformation.

Beauty: Training for a Targeted Market

Needs for Product Market Expansion

The collaborative training phenomenon is also prevalent in the service sector, particularly within the beauty industry. Companies here face significant challenges related to high labor turnover and skill shortages, motivating them to partner with vocational schools to cultivate a steady supply of ready-to-work employees. According to an internal survey by a research group from the Guangdong Department of Education, 83% of beauty firms in the local market encounter issues with labor turnover and skill gaps. To retain and attract talent, these employers often offer relatively high wages; the average salary for beauty workers was reported to be over 4000 yuan per month in 2015, compared to the local minimum wage range of 1210 to 1895 yuan across cities during the same period.

Moreover, leading companies in the beauty sector are motivated to train for targeted markets. These firms not only provide services through their own salons but also supply branded products (such as cosmetics and equipment) to franchises and small businesses. Consequently, they have a vested interest in transferring brand-specific knowledge and expertise throughout the industry to expand their market share. In an interview conducted in March 2016, the president of a leading beauty firm described their collaborative programs with schools as "incubators," aiming to train graduates not only as future employees but also potential entrepreneurs. By offering support in technology, expertise, and even capital, these companies ensure widespread use of their branded products and equipment across the market.

Many employers have shifted a significant portion of their firm-based training toward pre-employment skills development by establishing partnerships with vocational schools. The aforementioned report indicates that 77% of surveyed employers now hire through vocational schools, many under the "training for a targeted market" model.

At the core of this collaboration lies the objective of training students with firm-specific skills before recruiting some of them as formal employees. Schools benefit from enhanced employment opportunities for their graduates and access to cutting-edge knowledge and technology from firms. In return, they offer tailored curricula based on the specific needs of individual firms.

Mechanisms

The first strategy involves curriculum integration. A teacher dispatched by a beauty firm once explained how firm-specific technologies and skills are integrated into their training process:

> We follow modules provided by the firm and teach courses that can be immediately applied when students join our company. As part of the Taiwan Natural Beauty Group, we have a unique business model. Each firm in the beauty sector has different standards and processes... For example, we include training on a special skin detector—a machine I use prior to all other beauty care procedures—because I believe naked eyes can deceive. Another instance is body examination using iris scanning to detect past and potential health issues... Our teaching methods differ significantly from others, with some of our regular courses unheard of in other schools. (III: 5)

Their training plan includes 12 specialty-core courses, nine of which are named after the Natural Beauty Group or its brand, such as "Natural Beauty Health and Iris." These courses teach techniques and know-how specific to the services provided by the group in their salons, using textbooks also supplied by the firm.

Second, technology transfer plays a crucial role. Different programs utilize various firm-provided training materials, including internal training documents, and practical centers designed to mimic real workplace environments. An official from Qingyuan Polytechnic introduced their beauty practical center:

> Ellehuis Beauty invested heavily in our school, creating a practical center identical to their beauty salons. Students perform tasks almost identical to those in real salons, using the same equipment and following the same operational processes, including activating VIP cards. (II: 49)

Finally, knowledge transfer involves dispatching firm trainers. Partner firms sometimes send instructors to teach full-time at schools. In Guangdong's beauty industry, it is common for firms to assign a teacher to work as a full-time trainer at the school, focusing on application courses. During my visit to a beauty program, I met Ms. Xie, a full-time firm trainer who teaches approximately 60% of the program's specialty-core courses during the second year. She ensures students learn specific skills relevant to her firm. Typically, firm-dispatched trainers focus on application courses (e.g., beauty techniques), while schoolteachers handle theoretical courses (e.g., beauty basics).

Outcomes

This strategic collaboration between firms and vocational schools has yielded positive outcomes in terms of student employment and placement, receiving acclaim from both employers and vocational institutions. It aligns well with the central government's apprenticeship reforms, detailed documentation of which, along with a critical evaluation of their outcomes, is presented in Chap. 4.

Summary

Faced with the dual pressures of skills upgrading and the need for numerical labor flexibility, Chinese employers have increasingly turned to the pre-employment skills formation process as a strategic solution. Given the well-documented shortcomings of the vocational education system in delivering industry-relevant competencies (Cooke, 2005; Durden & Yang, 2006; Woronov, 2012; Song & Xu, 2024), this chapter reveals how employers have innovatively developed collaborative partnerships with vocational schools. This emerging model transcends the traditional labor agency depiction (Smith & Chan, 2015; Chan et al., 2015), in which schools merely function as recruitment pipelines, by incorporating meaningful employer involvement in curriculum design, training delivery, and technology transfer.

While low-level, transactional school-firm collaborations certainly persist across Chinese industries, I have observed a growing number of more strategic and targeted training initiatives that span a diverse range of sectors—including traditional manufacturing (e.g., construction), modern manufacturing, and service industries—as well as their overlapping domains. This study examines 21 firms and their collaborative programs with vocational schools across these sectors (see Fig. 3.1). Despite shared challenges such as skilled labor shortages, employer motivations vary significantly by industry, reflecting distinct concerns such as labor market instability, firm-specific skill gaps, high turnover rates, and bottlenecks in value-chain labor supply.

This grassroots model of strategic school-firm collaboration emerged well before the central government's formal apprenticeship reforms and

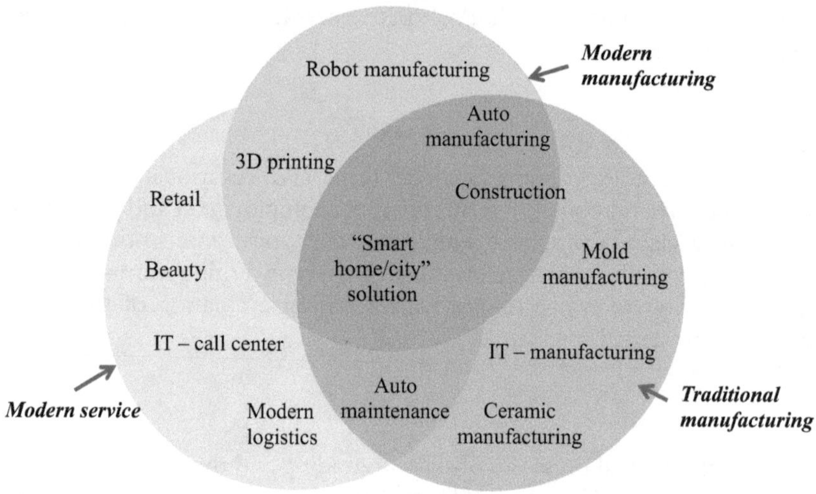

Fig. 3.1 Sectoral diffusion of collaborative training programs in China

may partially resolve the apparent paradox highlighted at the outset of this book and chapter: namely, the coexistence of systemic inefficiencies and the labor agency stereotype in Chinese vocational education, on one hand, and firms' persistent needs for skilled workers, on the other.

I therefore conceptualize the labor agency model and the strategic, integrated partnership examined here as two ends of a continuum. A critical question that follows is: under what conditions does the more effective, integrated form of school-firm collaboration emerge? Specifically, what motivates firms to make substantial investments—curricular control, technology transfer, knowledge sharing—in vocational education programs? In other words, how is the classic collective action problem (Streeck, 1989)—particularly the free-rider dilemma—overcome in a context lacking institutional mandates for such cooperation?

Several industrial and institutional factors help explain this phenomenon. While labor shortages are widespread across China's industrialized regions, strategic collaboration is more likely to develop in industries where workforce skills are highly specific. Skills specificity is closely tied to two key factors. First is the degree of technological specificity embedded in an industry's production or service processes. In cases presented above, it is the proprietary car models, robotic systems, branded cosmetics, and specialized skin-detection equipment that necessitate workers trained in firm-specific techniques. As a result, firms are strongly incentivized to externalize part of their internal training into the pre-employment education phase.

In contrast, industries such as electronics and garment manufacturing—characterized by deskilled, Taylorist production systems and pronounced seasonal fluctuations (Lüthje et al., 2013)—tend to rely on flexible, low-skilled labor pools, including temporary workers and student interns. In these contexts, collaboration with vocational schools, if it exists, typically remains confined to the labor agency model, with minimal investment in skill development.

Moreover, even within the same industry, certain occupations are more likely than others to trigger deep collaboration. A telling example comes from the automobile sector: maintenance technicians require a significantly higher level of specific technical knowledge compared to assembly-line workers. Consequently, branded automakers have stronger incentives to establish comprehensive training partnerships with vocational schools to ensure a steady supply of qualified maintenance personnel. While collaborative programs also exist for assembly-line roles, they tend to align more closely with the labor agency end of the spectrum.

Although employer engagement in vocational education is not unique to China, the Chinese case is distinctive in that such partnerships have emerged organically, driven by strategic employer interests rather than formal institutional incentives or mandates. However, this bottom-up model exhibits a clear selection bias—favoring industries and occupations with high skill specificity and strong brand or technological differentiation.

To generalize this successful model beyond its current scope—particularly to industries entrenched in low-skill, flexible labor strategies—formal institutional mechanisms must be established to enable effective cost-sharing among stakeholders: employers, schools, students, and even non-participating firms that benefit indirectly through labor poaching. Currently, poaching remains a dominant labor acquisition strategy in many sectors (Sheldon & Li, 2013), exacerbating the collective action problem first outlined in Chap. 1.

Drawing on both Western and Chinese experiences, deep collaboration tends to be a "big firm game": leading firms possess both the motivation and capacity to invest in training not only for their own needs but also for their clients, franchisees, and broader value chains—even absorbing some losses from sunk costs. Yet, for this model to catalyze structural transformation across the wider Chinese vocational education system, top-down policy interventions are essential. It is against this backdrop that the Chinese state launched two major apprenticeship reforms beginning in 2015, modeled in part on the German dual system. To these reforms I now turn in the following chapter.

CHAPTER 4

Institutionalization: Outcomes and Challenges of the Apprenticeship Training Reforms

The firm-driven, targeted training model discussed in the previous chapter has long operated in the shadows of China's vocational education system, tacitly accepted and at times even encouraged by state authorities. Its resemblance to the much-admired German apprenticeship system—known for its dual structure of school-based and workplace-based learning—has lent it a certain legitimacy in policy circles. By 2015, this informal alignment with international best practices culminated in formal recognition: the central state institutionalized the model through a nationwide apprenticeship reform, positioning it as a key component of the industrial upgrading agenda laid out in the "Made in China 2025" initiative. The underlying logic of these reforms was straightforward—to strengthen the supply of skilled labor by embedding training more deeply within actual production processes, thereby aligning workforce capabilities with the demands of a modernizing economy.

Both state agencies that have broad authority over skills development in China—the MOHRSS and the MOE—independently launched their own apprenticeship reform. However, while the MOHRSS has employed a state-led top-down approach to deliver apprenticeship training programs, the MOE has adopted a collaborative model involving non-state actors, including firms, vocational schools, and industrial associations. These vastly different approaches have led to considerable variations in their outcomes. For example, by mid-2016, the MOE had established ten times as

H. Zhang, *Upskilling the Chinese workforce?*,
https://doi.org/10.1007/978-981-95-2656-7_4

many apprenticeship programs as the MOHRSS had in Guangdong Province. Moreover, the former's programs had typically been welcomed by employers, whereas the latter's ones had triggered complaints from both employers and vocational schools.

To examine and account for this variation, in this study, I draw on the theoretical framework for evaluating the success of skills formation systems established in Chap. 1. In that framework, as summarized in Table 4.1, I contend that successful VET reforms should overcome three crucial obstacles in skills development: underinvestment in training, skills mismatches between curricula and industrial needs, and imbalance of skills formation, that is, the focus on either general skills or firm-specific skills, which have very limited industrial transferability. Specifically, in this chapter, I examine employers' willingness to participate as a key criterion for overcoming the underinvestment problem; I measure employer satisfaction as an indicator of addressing skills mismatches, that is, whether the training accommodates their actual needs, and compare the consistency of the curricula of all training programs within an industry to explore whether the programs achieve a balanced delivery of both transferable and firm-specific skills.

Using these criteria, I systematically compare the two models of apprenticeship reforms in China. I find that the MOE's collaborative model successfully engages firms to invest in the skills formation process as well as better accommodates industry needs by tailoring curricula and training processes based on employers' specific needs. By contrast, the top-down approach of the MOHRSS fails to solve these problems. However, the reforms of both the MOE and the MOHRSS fail to address the third issue, namely, the imbalance between general and firm-specific skills, in the absence of effective civil society governance in China. The MOHRSS's centralized apprenticeship system imposes standardized training based on the NVQS, leading to training programs that are overly focused on general skills. On the contrary, the MOE has granted individual programs full autonomy to develop their own curricula and training processes. This has resulted in training that is dominated by the skills development agendas of individual collaborating firms and, thus, frequently focuses on their specific skill needs without effective coordination across the industry.

Hence, I argue that the approach of the MOE is more likely to be successful than that of the MOHRSS because it involves collaboration with non-state actors in ways that address the underinvestment and skills mismatch problems. However, this model is not ideal either, as it may not

Table 4.1 Key problems in skill development and performance assessment of the apprenticeship reforms in China

Key problems in skills development	*Institutional resolutions*	*Assessment criteria*	*Indicators*	*MOHRSS: Top-down model*	*MOE: Collaborative model*
Underinvestment in skills development: skills are essentially public goods and have positive externalities	State intervention and subsidy—to oblige and/or incentivize cost-sharing among firms and individuals	Employer participation	Number of schools	4	19
			Number of programs	5	51
			Number of firms	5	51
			Number of students/apprentices	N/A[a]	2350
Skills mismatches: skill formation processes must be compatible with industrial development	School-firm collaboration—to engage and motivate firm participation	Employer satisfaction	Number of cycles	<1[a]	2
			Employer feedback	Wide complaints	Welcomed by employers
Imbalanced skills training: to avoid overly focusing on either general or individual firms' specific skill needs	Corporatist civil society governance—to widely coordinate firms' interests	Balanced skills training	Curricula consistency	Overly focused on general skills	Overly focused on firm-specific skills

[a] Apprentice admission not completed by mid-2016

meet the society's needs for a sustainable vocational training system that can balance general and firm-specific skills to achieve skills transferability. In the absence of effective civil society governance mechanisms to coordinate the interests of the industry, the Chinese skills development system is trapped in an institutional dilemma—it has either overly centralized reforms that fail to accommodate regional and industrial specificities or overly decentralized reforms that only cater to the immediate needs of individual firms. This institutional dilemma poses a major challenge to the Chinese state's long-term agenda of industrial restructuring through upgrading the vocational skills of its workforce.

Top-Down Reform of the MOHRSS

Following the central state's call for experiments with apprenticeship training in its "Made in China 2025" initiative, the two ministries each launched an apprenticeship reform. However, as previewed in Table 4.1, the outcomes of both reforms in Guangdong by mid-2016 were vastly different, with the MOE's reform being generally more successful than the MOHRSS's. This divergence is explained by the two ministries' distinctive approaches to the reforms. As summarized in Table 4.2, the MOHRSS followed a state-led top-down model that relied overwhelmingly on detailed policies and guidelines, quotas, subsidies, and a centralized skill

Table 4.2 Features of the two reform models and their performance in the three key problems

Key problems in skills development	*MOHRSS: Top-down model*	*MOE: Collaborative model*
Problem I: underinvestment in skills development	Centralized guidelines disqualify most firms; conflicting central and local policies discourage firm participation	Authority diffusion and integration of local experiments allow a wide scope of participation
Problem II: skills mismatches	Rigid policies fail to integrate firms' needs	Flexible arrangements allow training programs to be tailored for collaborative firms
Problem III: imbalanced skills training	NVQS-based training programs only focus on general skills	Single employer-based training agendas narrowly focus on a single firm's specific and instant skills needs

credentialing system to motivate employer participation and guide them through the training process. The MOE, by contrast, adopted a collaborative approach to apprenticeship training. With substantial authority diffusion, the ministry chose to integrate existing school-firm collaborative programs in local vocational schools into their reform project and grant them sufficient autonomy to develop apprenticeship programs based on their own idiosyncrasies. I examine the two models in turn, beginning with the MOHRSS and how it fails to resolve the three key skills development problems.

Lack of Employer Participation: Failure to Resolve Problem I

The MOHRSS launched its "New Apprenticeship" (*xinxing xuetuzhi* 新型学徒制) system in July 2015 to facilitate employer participation; however, ironically, the reform failed to motivate local firms in Guangdong to participate mainly because of its very rigid central policies. Its central policies offer a detailed guideline and roadmap that direct firms through the institution-building process, yet impose strict standards and quotas. In particular, only large- and medium-sized firms that have over 60% skilled workers in their workforce and an established internal training system are allowed to participate. In each cycle, a participant firm must recruit and enroll no fewer than 100 apprentices on a one- or two-year program. Participant firms, thus, purchase these subsidized programs from collaborative vocational schools to allow the apprentices to acquire a relevant national vocational certificate.

These standards essentially excluded many firms that were interested in state subsidies for vocational training, but did not meet all of the requirements. In particular, small firms and start-ups desperately need high-skilled workers, but lack the capacity to provide systematic internal training for staff. The Internet of Things Association of Guangdong Province, which has a membership of over 600 start-ups and established firms in the local IT industry, runs a training program with effective coordination among its member firms. Since its inception, it has self-funded this program and has long been eager for support from the local government; however, the MOHRSS's reforms excluded it from the project. In other words, the MOHRSS could have been more successful if it had sought to integrate such well-established programs into its own system instead of using a rigid state-led top-down model and only focusing on local "celebrity firms".

In addition, the MOHRSS provided subsidies with the aim to share costs and encourage firm participation, but their policies conflicted with existing local policies in Guangdong. While the ministry promised to cover 60% of the tuition fees, the Guangdong Department of Human Resources and Social Security (GDDOHRSS) was already subsidizing 100% of the same NVQS training for individuals that successfully completed the course. Therefore, state intervention, which was supposed to bring more employers on board, became a de facto disincentive to participation.

While many firms were excluded, the state had to use its administrative power to designate several local "celebrity firms" to enroll. However, these firms did not show enough enthusiasm for it. In October 2015, the GDDOHRSS formally announced five experimental apprenticeship programs with five firms and four partner technical schools as participants. Local state officials generally welcomed my fieldwork in Guangdong, partly because of my connections to central ministries, except for the entrée into these experimental programs. While I was fortunate to access one of them, the Toyota case, via my other networks, the unusual quiescence of GDDOHRSS officials about their apprenticeship reform was revealed by an inside informant:

> The GDDOHRSS issued a subsidy policy, but no firm has applied for the funding. The department director was very concerned. We communicated a couple weeks ago, and they asked what was wrong with their policy because no single firm had sent in their proposal so far…None of the five selected firms had applied for the money, and they were so anxious. (IV: 2)

Thus, as shown in Table 4.1, the MOHRSS's reforms were less successful in resolving Problem I compared with those of the MOE. It had recruited only five firms and four schools into five apprenticeship programs in its first cycle in Guangdong, and none of these programs had completed the apprentice admission process by mid-2016.

Participants' Complaints: Failure to Resolve Problem II

The rigid policies have triggered many complaints by participant firms. The Toyota case helps explain the complaints and reveal how the MOHRSS's model failed to resolve Problem II. Established in 2004, Toyota in China is a joint venture between the Guangzhou Automobile

Group (GAC), a state-owned enterprise, and Toyota (Japan). In 2016, the auto assembly firm employed 9600 workers and had an annual production capacity of 380,000 cars.[1] As Toyota holds a 50% share of the joint venture, the human resource management system of the firm largely follows a typical Japanese model that offers continuous internal training to workers. As a local "celebrity firm" that continues to receive generous support from the local state, Toyota was unsurprisingly selected to participate in the GDDOHRSS apprenticeship reform in 2015.

However, Toyota could not overcome two major obstacles that prevented its whole-hearted participation in the reform. The first was the conflicting central and local policies regarding subsidies, as noted. Now that Toyota had budgeted approximately 100,000 yuan for its first cycle of apprenticeship training, based on its own estimation, it had to incur the unnecessary cost of approximately 45,000 yuan. Second, Toyota was angered by the strict policy requirement of offering 100 apprenticeship positions, which had to be newly recruited workers, in a cycle where they did not even have these many positions open to skilled workers in a year, let alone inexperienced apprentices. Thus, Toyota circumvented the policy by enrolling 100 existing workers on the training program.

The MOHRSS's apprenticeship reform triggered many complaints from Toyota. From employers' perspectives, such national vocational certificate-based trainings were useless. A pseudonymous human resources specialist from Toyota, Ms. Wang, was appointed to coordinate the apprenticeship program in collaboration with the Guangzhou Technician College as per the GDDOHRSS's requirement. It is worth quoting a section of my interview with her:

> Author: Why did Toyota decide to do this?
>
> Wang: It is required by the GDDOHRSS. Their commanding document was sent to us… We said in private that this is a task from the state, so we cannot really resist it. We also have a general manager from the Chinese partner—we are attached to the GAC, a state-owned firm. So, we have to participate whether we are willing or not.
>
> Author: What do you think about the future of this program?
>
> Wang: (Shaking head) In the long run, we will definitely suffer a loss. Even if the subsidy is increased to 100 per cent in the future, it is still a big

[1] Data were collected in October 2016 from Toyota's official website, available at: http://about.gac-toyota.com.cn/visit/index.html. The figures cited may have changed since, as the actual numbers are subject to periodic updates.

> burden for us. Yes, we would encourage our workers to go and get these certificates...But if you ask us to arrange for them to do so, [we are not very willing] because these certificates are not very useful for us. We have our internal training system that is based on our own technologies.
>
> Author: Are relevant NVQS certificates recognized in the auto industry in general?
>
> Wang: We are happy if workers have these certificates, but they are not necessary. For instance, the certificate for auto maintenance is not really linked to real workplace [situations].

Toyota enrolled on the program because they viewed it as a requirement from the state, but was reluctant to turn in their already finalized proposal to the GDDOHRSS. Indeed, none of the selected experimental firms had done so by mid-2016—nine months later than the MOHRSS's roadmap required it.

All in all, the MOHRSS's reforms triggered numerous complaints from employers, leading it to be less successful in resolving Problem II compared with the MOE's programs, which received positive feedback from employers for accommodating industry needs.

General Skills-Focused Training: Failure to Resolve Problem III

The human resources specialist's comments above revealed an important fact: these NVQS-based skills trainings only focused on general skills, thus, employers often found them useless. Established in the 1950s, the NVQS was initially used to manage the employment and compensation of state workers in the centrally planned economy. Until the market economy reform in the 1970s, a worker's wages and benefits were linked to the NVQS. The economic reform sought to break the "iron rice bowl" of job security and grant firms the discretion to determine their own wages and benefits. The NVQS has since become increasingly incompatible with industry needs, and NVQS-based trainings have thus become unpopular among Chinese employers. The reform therefore failed to resolve Problem III and achieve a balanced delivery of general and firm-specific skills.

The NVQS-based trainings have received a variety of criticisms. First, the credentialing system is too centralized to accommodate industrial and regional specificities. The MOHRSS exclusively set the skills standards and make the curriculum of the NVQS,[2] while local Bureaus of Human Resources and Social Security and their approved entities—typically public technical schools—only carry out the qualification and training process. Many interviewees complained about the backwardness of the NVQS; it cannot catch up with the relatively advanced industrial development in Guangdong. A school provost provided a good example:

> We went to a food factory, and set up an inspection laboratory for them…Every food factory is required to employ a fulltime food inspector, but workers [holding a food inspector certificate] often cannot really get things done. A vocational certificate does not attest to the actual skill level of a worker…The MOHRSS had a number of old experts make up the skills standards, and had not updated it for years. These experts are probably unfamiliar with the industry, so the qualification does not really test the most important things. (II: 49)

Second, major updates of most skills standards occur every five years, rendering these standards outdated and unable to follow the rapid industrial development. For instance, in VW's new Foshan factory, I saw welders using robotic machines for welding work. A workshop manager relayed that they had no interest in hiring a welder with a national certificate that only attests to his/her ability to operate traditional manual welding guns.

Finally, a deeper institutional factor that accounts for the failure of the NVQS is the absence of effective coordination among employers at the industry level, which could potentially resolve the problem of imbalanced skills transferability. Urged by Prime Minister Li Keqiang, the MOHRSS has recently aimed to reform the NVQS, and one of the strategies is to let industrial associations run the relevant vocational certificates. According to my interview with an MOHRSS official, however, the ministry is reluctant to authorize these civil society actors because they are unconfident about their capacity:

> Employer associations in China are underdeveloped…Traditional industrial associations are not real employer associations, but rather former govern-

[2] In some exceptions, industry associations or industry-relevant state ministries are authorized to manage certain professional certificates and related qualification processes, such as doctors and lawyers. These certificates are generally more favorable in relevant industries.

> ment institutions, whereas those associations self-organized by firms do not have adequate capacity [to run vocational certificates]...We certainly hope to maximize their role, but not until they can represent their industries and govern their memberships. (I: 2)

In Guangdong, evidence largely supports this argument. A director of the official local chamber, the Guangdong Provincial Enterprise Confederation, relayed to me:

> To be honest, our confederation is doing nothing with regard to education and training. (Author: Why?) We lack the capacity, and enough employees, and we cannot understand what firms need. We used to have a department for training, which was closed later because they could not develop any profitable business. (IV: 15)

Therefore, skills standards setting of the NVQS is likely to remain centralized in the long term. Now that the MOHRSS applies the NVQS to its apprenticeship system, a fundamental concern is that curricula derived from these standards would fail to cater to industry specificities. The Toyota program was composed of three separate classes focused on the middle- and senior-level certificate of auto maintenance, and the middle-level certificate of auto examination. The director of the Guangzhou Technician College's training center, Ms. Chen, relayed that the training would strictly follow the relevant NVQS standards (i.e., focus exclusively on general skills training). In fact, I did not identify any module in their training plan that was tailored to Toyota, indicating that the plan focused on general skills training only.

Collaborative Model of the MOE

The MOE initiated its "Modern Apprenticeship" (*xiandai xuetuzhi* 现代学徒制) reform in August 2014, and announced a list of experimental programs a year later, only two months before the GDDOHRSS. Due to their collaborative model that was ready to engage local players and incorporate their existing experiments with various formats of school-firm collaboration, their reform in Guangdong by mid-2016 turned out to be more satisfactory.

Integration of Local Experimental Programs: Successful Resolution of Problem I

Different from the MOHRSS, the MOE adopted a decentralized, collaborative approach to its reform, encouraging schools and firms to actively contribute to the reform process rather than merely follow state-issued guidelines. The MOE's August 2015 "Notification" on apprenticeship reform stipulated only four general principles and asked experimental programs to establish and publish their own roadmaps and guidelines in coordination with local education departments. This allowed firms and training institutes greater autonomy to define their own programs based on their individual circumstances.

In addition, the MOE's subsidy policy was flexible. While the MOHRSS required firms to make a detailed trainee-based budget and promised to reimburse 60% of the costs afterwards, the MOE disbursed a lump-sum subsidy of 500,000 yuan (for the seven programs in the first cycle and 100,000 yuan for the later programs) to each program beforehand, without requiring them to provide detailed budgets. Moreover, those conducting the program had full autonomy to use this money to cover any costs associated with the training. Although the majority of the interviewed program directors claimed that this was insufficient to cover all costs, they did appreciate the flexibility.

The flexible arrangements of this model have allowed many existing local experiments with school-firm collaborative training to be integrated into the state's formal reform agenda. As presented in Chap. 3, Local vocational schools had continuously experimented with the collaborative, targeted training. Employers sponsoring these classes sought to shift part of their internal training into the pre-employment skills formation process for prospective workers, and thereby gained an advantage over other firms when selecting graduates. Such collaborative training was typically delivered at the individual school-firm level, in contrast to western apprenticeship systems that coordinate training content and skills standards across employers at the industry level.

Therefore, the MOE's model can engage employers of a much wider scope within an industry, and thus, better resolve the underinvestment problem of skills development. The MOE's apprenticeship system generated two cycles of 51 programs in 19 schools; my research covered 9 schools, including all the 7 involved in the first cycle of apprenticeship experimentation.

Participants' Satisfaction: Successful Resolution of Problem II

The flexible arrangements triggered extensive collaboration between vocational schools and employers and won their satisfaction. Here, we use another multinational carmaker, Jaguar-Land Rover, and their program, to further demonstrate the process. The Guangdong Mechanical and Electrical College has engaged in education and training for the auto sector for over 50 years, maintaining collaborations with VW, Jaguar-Land Rover, and other major carmakers over time. Its collaborative program with Jaguar-Land Rover since 2012 involves a co-built practical center, curriculum and textbook development, teacher training, and firm-dispatched trainers that work full-time in the school, among other activities. In 2015, this "customized class" became an apprenticeship program, one of the MOE's seven experimental programs in Guangdong in the first cycle.

The curriculum was tailored to the skills needs of Jaguar-Land Rover. A class of 30 apprentices was recruited in the same year to enroll in a three-year training process, as follows: classroom-based general skills training of 900 credit hours in the first year, on-campus practical center-based firm-specific skills training of 486 hours in the second year, and firm-based job-specific skills training of 1020 hours (i.e., apprenticeships) in the final year.

This customized training process apparently satisfied all the participants. The apprentices that I interviewed generally expressed their satisfaction with the program. When asked what he appreciated most about this program, one apprentice said:

> In our first year, we studied a wide range of things. So, when it comes to the second year, we must focus on a specific brand's car models. We should neither seemingly know many cars nor be unfamiliar with any. We have to be specific if we want to be an expert. This program provides such an opportunity, and has rich content. It gives us a deeper understanding of a specific brand. (II: 107–114)

Jaguar-Land Rover's management also evaluated this collaboration positively and, in 2016, recruited its second cycle of apprentices. In addition, every year, the firm invites schoolteachers onto three or four training tours at the Jaguar-Land Rover headquarters in Shanghai as well as facilitates regular visits to its 4S stores in China to update their skills.

Single Employer-Focused Training: Failure to Resolve Problem III

This model that seemingly satisfied all participants may undermine the local labor market in the long run because of its failure to resolve Problem III, that is, to achieve a balanced delivery of general and specific skills. The collaborative training is exclusively institutionalized at the individual school-firm level. As schools enjoy full autonomy to develop their own curricula under the MOE's apprenticeship reform, the training tends to focus on firm-specific skills. While the Jaguar-Land Rover program has somehow managed to retain general skills training for a year, this may not be the case in other industrial settings.

To estimate the skills transferability of the collaborative programs across the local industry, I coded and compared the curricula of all ten beauty programs from schools within the Guangdong Health Vocational Education Association (GHVEA). Each of these programs collaborated with a beauty firm; three had engaged in apprenticeship training, including one within the MOE's reform. As shown in Fig. 4.1, all these curricula

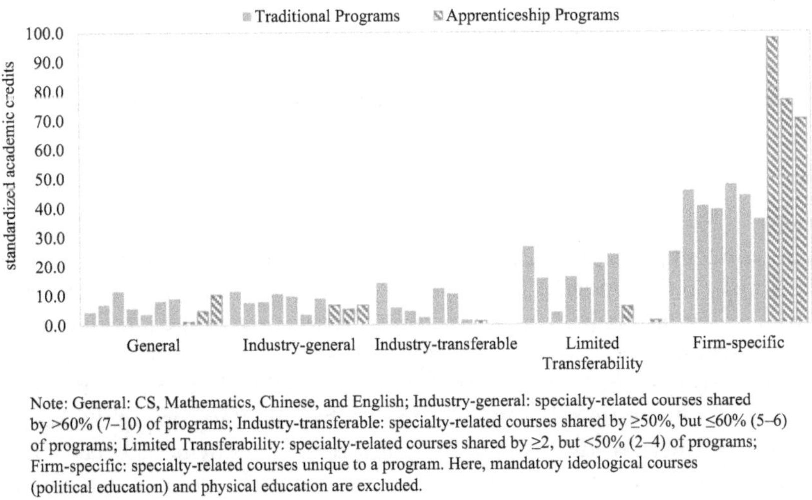

Fig. 4.1 Distribution of the training modules (academic credits) of the ten major training programs in Guangdong's beauty industry across five levels of skills transferability

tended to focus on the specific skill needs of collaborating firms. The three apprenticeship programs highlighted are even more firm-specific than the other seven ordinary vocational education programs.

This inconsistency of curricula across training programs for the same specialty has led to incoherent skill structures for the workforce within the local industry. In 2016, the GHVEA conducted a provincial skill competition in the beauty industry. I participated in their preparation process as an external observer and noted tremendous controversy among schools regarding what skills are essential to a beauty worker's job and thus must be tested in the competition. During a meeting in January 2016, for example, the directors of member schools engaged in drastic debates regarding the content and qualification standards required for the skill competition. I was astonished that they could not even reach a consensus regarding issues like the basic treatment techniques of the Chinese traditional massage. On this specific issue, a program teacher from the Zhuhai Health School spoke against other programs. She argued:

> Our massage training focuses on the know-how in the lymphatic drainage service in aromatherapy. So, we do not teach this 'rubbing' treatment. The lymphatic drainage is to diffuse the lymph to the armpits, so we use only 'pushing', but not 'rubbing'. (Another teacher: Why do you not teach students all the basic techniques and then focus on the use [in different services]?) That is because we do not offer those services in our salons. Our training is targeted [for our collaborative firm], so we do not want to teach them many irrelevant things (Field notes, February 2016).

Considering this, even though firms and schools, and frequently students themselves, are quite satisfied with these single employer-focused apprenticeship programs, the negative effects can be significant. Overly focusing on firm-specific skills in workforce training not only undermines workers' employability and self-learning capacity across the local industry, and therefore labor market mobility, but also increases retraining costs of firms, further aggravating the skilled labor shortage in China. While systematically estimating the level of impact is beyond the scope of this study, there are hints already in the beauty industry. An anonymous beauty program director commented on the Zhuhai Health School's over-focus on Natural Beauty's specific skills:

> I feel their focus is too narrow. They (the firm) have iris detectors, so they (the school) offer an iris examination course. But this technology is only used in Natural Beauty. If I do not work there, it is useless... The curricula developed based on one firm's practice, I can by no means apply it to another one. They are completely incompatible. (Field notes, May 2016)

An interviewed beauty worker said:

> Although I did not do well in my Chinese Medicine Basics course, (I took it at least). After I entered the workplace, every salon has different trainings for us because they have different products and different massage technics... This cannot be changed, but I feel it is very easy for me to understand those trainings. Many colleagues from other programs, they do not even know where each acupuncture point is located. (II: 78)

This indicates that "training for a target brand" may be even more problematic given that it is within the degree training programs, which are supposed to grant students adequate general skills (e.g., the Chinese Medicine Basics course mentioned by this worker) that are indispensable for their future continuing education and self-learning processes.

Various forms of labor organizations in Germany, namely, labor unions and works councils, assume important responsibilities in workers' skills development processes at both industry and organizational levels. This functions as an important mechanism to prevent training processes from being overly focused on firm specificities. By contrast, official unions in China, the ACFTU and its firm-level union committees play only a limited role, or have no role at all. According to a former union official in Guangzhou, the only function that unions perform in workers' skills development nowadays is organizing skill competitions within firms (IV: 11).

While employers and labor organizations lack interest and necessary capacity, formal and informal school networks can occasionally serve as coordination platforms. In the beauty industry, for instance, the GHVEA is where major beauty training programs in Guangdong form coalitions and coordinate on curriculum development and standard making at the industry level. Such cases are not commonly observed in the other sectors. However, even in the beauty industry, the GHVEA lacks governability over its membership. The chairwoman, Ms. Li, is a retired former health-care school president. She is now leading a group of four full-time workers in the association. She regarded the GHVEA as a "service platform" that lacks governability over its 32-member schools and 6-member firms where "even paying member fees is voluntary" (IV: 7). An active director of a member school commented about the association as follows:

> As a platform and communication mechanism that connects schools and the state, and promotes skill competitions, the association plays some role… But, commitment needs clear rights and obligations. What can they give us? If we do not obey their standards, what sanctions can they impose on us? Nothing. My boss told me: whatever the association tells you to do, if useful for our specialty development and consistent with what you think, you listen; Otherwise just ignore it. (II: 65)

Clearly, it is members' strategic cooperation, instead of any compulsory mechanism, that brings members together in the GHVEA. Therefore, it is doubtful if the association can impose industry-level curriculum standards on its member schools, as doing so obviously conflicts with members' self-interests in serving individual collaborative firms' specific training agendas.

Overall, the MOE's reform has thus failed to address Problem III and cannot grant sufficient skills transferability to workers.

All in all, both MOHRSS and MOE's reforms have failed to secure a balanced delivery of general and specific skills. The MOHRSS's programs offered general skills training, whereas the MOE's were overly focused on firm-specific skills.

Summary

With the central state recently shifting its focal point of reforms toward supply-side labor market policies, both the MOHRSS and the MOE have embarked on tremendous efforts to reform vocational schools within their jurisdictions by establishing an apprenticeship system under the overall agenda of upskilling the Chinese workforce. However, both the data and my research show that the MOE's reforms had achieved more favorable outcomes in terms of the scale and level of satisfaction of participants than the MOHRSS had by mid-2016. What explains this divergence, and what are the implications for understanding the Chinese state and its economic reform in general?

This divergence lies in the different reform strategies of the ministries. The MOE's collaborative model has an advantage over the MOHRSS's top-down approach for addressing key theoretical problems in skills development. First, both approaches have the potential to solve the underinvestment issue and encourage firms to offer apprenticeship positions, but the MOHRSS's top-down model has inherent problems. Chinese reforms have often relied on local experimentations, and the successful ones have the potential to be rolled out nationally (Friedman & Kuruvilla, 2015). Thus, state-initiated collaboration has provided a valuable opportunity for integrating some of these important local experiments into the central reform agenda. The top-down model of the MOHRSS excluded these local experiments at the policy level, while the collaborative model of the MOE embraced them, leading to the latter's success. The central policies of the MOHRSS also conflicted with local ones, becoming a de facto barrier to firm participation in Guangdong.

Regarding the long-existing skills mismatches in the Chinese VET system (Cooke, 2005; Durden & Yang, 2006), my findings suggest that effective school-firm collaborations will generate better outcomes as it better caters to employer needs. The MOE's apprenticeship programs have surpassed the MOHRSS's in terms of employer participation and satisfaction as well as program sustainability because of schools' abilities to tailor their curricula to employer's needs. The MOHRSS's reform does not allow that flexibility. Indeed, it requires firms to purchase training based on the centrally standardized NVQS, which has proven ineffective in meeting regional and industrial needs. The complaints from employers and the local government demonstrate the rigidity of this centralized approach.

Lastly, both the MOE and the MOHRSS have failed to address the problem of imbalanced skills transferability. The top-down model of the MOHRSS excessively focuses on general skills, partly explaining why it often fails to satisfy employers. The MOE's collaborative model is faced with a seemingly insurmountable obstacle, namely, the absence of civil society governance in China that is crucial to exemplar national models succeeding elsewhere. Instead, the MOE's apprenticeship programs are based on collaborations between individual employers and schools that lack effective coordination at the industry level, rendering them overly focused on the specific skill needs of collaborating firms. These challenges are likely to persist, as highlighted by a more recent study (Müller, 2024). The findings therefore carry important implications for understanding and advancing current skills development reforms in China—reforms that will be the focus of Chap. 5.

Before turning to the final chapter, it is important to recognize that while the MOHRSS's apprenticeship reform has followed a top-down, centralized model, this does not reflect the full picture of its training programs. Across the broader network of technical schools under its jurisdiction, the delivery of vocational training remains highly decentralized. In fact, the targeted brand model of training identified in Chap. 3—where schools align their programs closely with specific enterprises or well-known brands—is deeply embedded in the Chinese vocational education system and is just as prevalent in MOHRSS-affiliated technical schools as it is in those overseen by the MOE. This widespread practice underscores the complexity of China's vocational training landscape, where centralized policy frameworks coexist with locally adaptive, market-responsive models shaped by institutional legacies, local industrial demands, and school-firm relationships.

CHAPTER 5

Dilemma: The Chinese Reform in a Comparative Institutional Perspective

In this book, I argue that in its efforts to reform the vocational education system during the past decades, the Chinese state has become trapped in a dilemma between centralization and excessive decentralization. While the state has aimed to upskill the Chinese workforce and sustain the "economic miracle" into the forthcoming decades, it is attempting to accomplish this without including effective civil society governance in the process. This has posed challenges to its reforms, as neither a centralized nor a decentralized approach is likely to deliver ideal skill formation outcomes.

Specifically, I have delineated three major stages in the evolution of the Chinese vocational education system since the economic reform and opening-up: decentralization, collaboration, and institutionalization.[1] The state initiated a major decentralization reform in the mid-1980s. As a result, state authority has largely withdrawn from the daily operations of vocational schools, and schools have gained nearly full

[1] To clarify, my aim here is not to establish a definitive historical periodization of the Chinese VET system—for such comprehensive historical accounts, readers are referred to existing scholarship (Wang, 2021; Xue & Li, 2022). Rather, my delineation of three developmental stages is analytically driven by the central narrative of this study: how the state has gradually withdrawn its direct authority from the VET system, how employers have increasingly stepped in, and how grassroots innovations have ultimately been institutionalized through state policy.

H. Zhang, *Upskilling the Chinese workforce?*,
https://doi.org/10.1007/978-981-95-2656-7_5

autonomy in specialty establishment, student admission, and curriculum development. This has resulted in a variety of vocational education models with different levels of training effectiveness and funding sources, generating graduates with distinctive skill levels and job market prospects.

This decentralization reform laid an important institutional foundation for collaborative training. Chinese employers, however, were not motivated to wholeheartedly embrace vocational schools to acquire skilled workers until the turn of the century, in the face of a major secondary labor market failure and shortage of skilled workers. Consequently, employers have gradually stepped in and taken over the training processes in at least a significant proportion of China's vocational schools, shifting part of their traditionally firm-based training into the vocational education process. The "training for a targeted brand" model has thus emerged from the bottom up. Firms may have varied incentives across different industries, and thus this model has variants—sometimes featuring customized training for an entire value chain of a lead firm or serving an expansion strategy aimed at a particular product market. But successfully implemented programs under this model share three important mechanisms: employer control over the curriculum, and sufficient knowledge and technology transfer from the firm to the school.

Without industry-level coordination or binding standards imposed by the state or unions, skill development under the "training for a targeted brand" model is entirely tailored, focusing on the specific skills required by a particular brand or firm. Given the absence of state regulation regarding a reasonable cost-sharing structure among firms, schools, the state, and individual trainees, it remains unclear whether public schools are justified in engaging in this kind of targeted training for a single firm within their degree programs. Nonetheless, this practice has become widely prevalent and tacitly approved by the state. Both the MOE and the MOHRSS have been promoting such school-firm collaboration, and many schools strongly encourage programs to find their "partners," regarding this as an important criterion for teachers' and department officials' performance evaluation.

This grassroots experimentation in collaborative training had developed in a largely disorganized manner until the central state initiated the apprenticeship reforms in 2015. This reform, however, revealed a drastic contrast between the two ministries' institution-building approaches. The

MOE has continued with the relatively flexible, bottom-up approach, offering full autonomy to vocational schools for experimentation, and institutionalizing and further decentralizing their collaborative and targeted training programs. By contrast, the MOHRSS has carried out its reform in a top-down manner, seeking to recentralize the system through strengthened enforcement of its NVQS.

In a systematic and critical assessment of these two separate reforms, I found that the centralization model of the MOHRSS has encountered major resistance from firms. It has therefore failed to attract sufficient employer participation, align training curricula with actual industrial needs, and ultimately deliver skills with firm and sectoral specificity. By contrast, the decentralized model of the MOE has institutionalized existing collaborative and targeted training programs in vocational schools and thus performed relatively well on these criteria. But it tends to overly focus on individual collaborative firms' specific and short-term skill needs, thereby undermining workers' long-term employability and labor market mobility.

Looking back over the last decade since the "Made in China 2025" initiative, the decentralized model and targeted apprenticeship training may have effectively alleviated the skilled labor shortage and sustained the competitiveness of the Chinese industrial sector in the global market. I argue, however, that the ultimate success of these vocational education reforms hinges on whether industry-level coordination can be effectively institutionalized—a key distinction between the Chinese and German models lies in what I term the coordination structure. While the German model has legally enshrined the partnership between industrial unions and employer associations, the Chinese model has been largely institutionalized at the individual school-firm level. As such, although the training programs appear superficially similar to the German model, the distinct and highly decentralized coordination structure has rendered Chinese apprenticeship training overly focused on individual firms' immediate, specific skill needs, while overlooking general skills training. In the absence of industry-level coordination across programs, I predict that this will eventually undermine the upskilling opportunities of the Chinese workforce and further aggravate the skilled labor shortage that stands in the way of Chinese industrial restructuring and upgrading.

Policy Implications for China's Vocational Education Reform

These findings hold important implications for the ongoing Chinese vocational education reform. The skill formation dilemma demonstrates that neither a top-down nor an over-decentralized approach to vocational education is ideal. To more successfully pursue long-term upskilling of the Chinese workforce, given the ongoing economic downturn of the country, the state should encourage the development of a robust coordination structure at the regional-industrial level. This coordination structure should include at least four functional components: a mandatory cost-sharing mechanism, a coordinated skill credentialing system, an internal governance structure composed of representative organizations, and an effective labor-capital dialogue mechanism. In particular, the state should focus on capacity building of civil society actors, primarily employer associations and labor unions, to promote civil society governance in the vocational education system. It should seek to activate the role of official employer associations, strengthening their organizational and governance capacity over members through legislation. It should also enhance the ACFTU's grassroots representativeness and its role in skill development.

The ideal skill formation scenario entails civil society governance to coordinate the diversified interests across firms and establish a long-term workforce training agenda at the industry level, with a balanced delivery of general and specific skills. Mechanisms should therefore be established at the industry level to coordinate the development of skill standards under the NVQS, reinstating it as the core of the actual training process in vocational schools and serving as effective benchmarks for program design. The state should also provide an effective institutional framework to coordinate mandatory cost-sharing among various stakeholders in the skill formation process, including the state, schools, collaborative firms, labor unions, employer associations, lead firms, NGOs, and students and their families. The state could consider activating the National Tripartite Mechanism for Coordinating Labor Relations as an important institutional platform for coordinating cost-sharing and skill formation agendas—not only within the vocational education system, but also linking them to wage policies and broader macroeconomic adjustment issues at the central and regional levels.

In light of the radically changing state-civil society relationships under the current regime of labor (Howell & Pringle, 2019), however, scholars have maintained relatively pessimistic attitudes toward such structural changes in China (Kuruvilla, 2018). The ACFTU and official employer associations may have relatively strong influence over the central state's policy-making processes, but often lack the capacity to mobilize their members at the grassroots level (Friedman, 2014a). Institutional frameworks developed to facilitate labor-capital coordination at more centralized levels, such as within regional industries, have been notably constrained (Kuruvilla & Zhang, 2016). For a fairly long period, therefore, the Chinese vocational education system may remain trapped in the skill development dilemma: vocational training programs will either overly focus on general skills by strictly following outdated NVQS standards—thus failing to accommodate industry and firm specificities—or overly focus on a single employer's specific and immediate needs while overlooking students' long-term employability in the broader labor market.

Theoretical Implications: China and the Comparative Literature on National Skill Formation Systems

Finally, the Chinese vocational education model holds important theoretical implications for the comparative institutional literature on national skill formation systems. While Chap. 1 has provided a review of that literature, the rest of this book returns to it and elaborates on how the China case enriches that literature from a comparative perspective.

Comparative scholarship on skill formation systems widely adopts a national system perspective (see Ashton et al., 2000 for a review). Scholars focus on characterization, typologization, and comparison of national skill formation systems. One prominent typological approach classifies national systems according to the main provider of vocational skills. These studies identify several types, including school-based systems in the United States, Japan, Canada, France, Sweden, and Italy; employer-led or dual models in Germany, Switzerland, and Austria; and mixed models, such as that in the United Kingdom (Furth, 1985; Green, 1991). Exemplary Western and Asian national systems demonstrate distinctive models of school-firm interplay and varying degrees of decentralization. Identifying common themes across these models highlights the peculiarity of the school-based yet employer-dominated Chinese system.

Western scholarship on skill development indicates that in both school-based and employer-led models, a certain level of decentralization and employer input into the vocational education process is indispensable. Different systems adopt distinctive approaches to decentralization, and the process is not unique to China. In Japan and Singapore, school-based systems are supplemented by extensive employer-provided workplace training. The German employer-led system and the United Kingdom's mixed model encourage direct employer involvement in the pre-employment skill formation process.

I find, however, that the Chinese vocational education approach to engaging employers is particularly decentralized and disorganized. It features extensive school-firm collaboration institutionalized at the level of individual organizations, with no industry-level coordination among schools, employers, or trainees. This institutional disorder could have been mitigated had China followed Singapore's example by instituting strong state intervention or state-organized employer coordination. Instead, through the decentralization reform, both central and local governments have granted individual schools and employers full autonomy to develop their own training agendas. The NVQS has long been criticized and marginalized from vocational education processes because it has failed to accommodate regional and industrial diversity. Private governance mechanisms that could establish skill standards at the regional industry level would be a superior solution, but such mechanisms are currently impractical due to the scarcity of proactive employer associations in China. The key institutional arrangements that underpin successful Western and Asian skill development models are largely absent in contemporary China, and its system therefore suffers from fragmentation and disorder.

I contend that, given the China case, two important perspectives have been largely overlooked in the national skill formation system literature. First, the national system approach—by reducing national complexity to a single, abstract model—downplays regional, sectoral, and occupational idiosyncrasies within each country (Crouch et al., 1999), thereby overlooking potential subnational variation. Locke (1992) suggests that subnational system analysis may offer a deeper understanding of industrial relations dynamics. Particularly relevant to skill formation, Osterman (1982, 1987) identifies four employment subsystems—industrial, salaried, craft, and secondary—each with its own distinctive model of skill development and HR strategies. This subnational lens is especially pertinent to the Chinese context, where vast regional disparities often explain critical variations in employment relations outcomes (Hurst, 2009; Friedman, 2014b).

The second underemphasized perspective is that of informal institutions. National skill formation system analysis tends to focus on formal institutions while neglecting informal ones. Helmke and Levitsky (2004) define informal institutions as socially shared rules that are mostly unwritten and exist outside official sanctions. Nee and Ingram (1998) remind institutional scholars that individuals and organizations are deeply embedded in social norms and informal networks, which shape many organizational processes. K. Tsai (2006), Tsai (2002, 2007), and Nee and Opper (2012) demonstrate that in China, informal institutions play crucial roles across socio-economic domains. Yet, research in the skill development field has largely failed to examine their influence.

Consistent with broader trends in Western scholarship, studies of the Chinese vocational education system predominantly adopt a national system perspective—proposing generalizations about the entire system while overlooking subnational variation and focusing exclusively on formal institutional structures. First, literature once portrayed the Chinese vocational education system as a school-centered model with very limited employer participation and input (Child, 1994; Cooke, 2005; Warner, 1993; Xiao & Tsang, 2004). Schools—both public and a small number of private institutions—were seen as the primary providers of initial vocational skills and constitute the core of the Chinese skill formation system (Cooke, 2005). Employer-provided, firm-based training was considered limited, concentrated in foreign-owned firms, joint ventures, and occasionally large SOEs (Wang, 2014), and primarily targeted at managerial and high-level personnel (Child, 1994; Warner, 1993; Xiao & Tsang, 2004).

A second key stereotype in the literature is strong state control, which was said to have been largely immune to the decentralization reforms of the mid-1980s (Lai & Lo, 2006). The state remained the largest provider of initial vocational skills. Most vocational schools were state-owned and primarily funded by state educational expenditures (Cooke, 2005). Although the 1980s reforms aimed to diversify funding sources by delegating financial responsibilities to local governments and introducing self-financing and social funding mechanisms, the central government continued to play a dominant role and remains a major funder (Schnarr et al., 2008). In the early 2000s, state policies shifted toward recentralization of both financing and supervision of education funding (Murphy & Johnson, 2009). Regarding curriculum development, specialty committees organized by local educational departments retained authority to approve new specialties in vocational schools, meaning schools themselves lacked decision-making power in this area (Lai & Lo, 2006). The MOHRSS and the MOE jointly control the graduate qualification system: the former managed the centralized NVQS, while the latter issued degree certificates. In sum, the decentralization in China's vocational education system had been limited by around the 2010s in literature. While some responsibilities were transferred to local governments during reform, the central government continued to tightly control the system through national policies and directives. Local governments were largely tasked with enforcement, and schools and institutions had little autonomy to design their own programs. That the state maintained strong control was not surprising—the ideological function of education has been crucial to a post-socialist regime (Althusser, 2006). Yet, this state centralization introduced rigidity into the vocational education process (Durden & Yang, 2006), which ought to be flexible enough to respond to rapid technological change in industry.

A third critical perspective in the literature depicts Chinese vocational schools as quasi-labor agencies. This view rests on two main arguments. First, the training quality and effectiveness of Chinese vocational schools were generally low (Song & Xu, 2024; Woronov, 2016). Many schools failed to equip students with useful vocational skills. Instead, they engaged in what Woronov (2012) calls a "mimetic labor" process—one that constructed students' social subjectivity as employable workers but rarely imparted genuinely valuable skills. Second, vocational schools were accused of selling student interns to businesses for unjustified commissions. Sociologists involved in a major research project on Foxconn exposed the abuse of student interns in Foxconn factories. Student workers were "forced" into low-skilled assembly-line jobs and assigned seasonal labor tasks unrelated to their specialties (Smith & Chan, 2015; Song, 2024; Su, 2010–2011). Given their ineffective training and labor brokerage behaviors, scholars concluded that Chinese vocational schools functioned as quasi-labor agencies. Smith and Chan et al. (2015) characterized such internship work as a form of constrained labor, while Su (2010–2011) referred to vocational schools as "middlemen."

Contrary to the prevailing scholarly consensus summarized above, my research reveals that the Chinese vocational education system is, as of now, neither centralized nor uniformly state-directed, but instead exhibits a decentralized and loosely coordinated structure, characterized by three distinctive features: (1) substantial institutional autonomy granted to individual schools in designing training programs; (2) significant inter-school variation that transcends the widely cited "quasi-labor agency" stereotype; and (3) a dominant pattern of employer-led, school-firm collaborative training at the micro level.

The first feature underscores a notable retreat of the state from direct involvement in the day-to-day operations of vocational institutions. While the central and regional governments continue to play strategic roles—such as setting broad policy directions and disbursing financial resources—they have largely ceded control over curriculum development and pedagogical practices to individual schools. This devolution of authority runs counter to the conventional view that the Chinese vocational system is tightly controlled by the state. Even the recent apprenticeship reform launched by the MOHRSS, which aims to recentralize program design through top-down mandates, has so far yielded limited institutional change on the ground. The persistence of local autonomy suggests that formal policy initiatives may struggle to override entrenched patterns of decentralized implementation.

Second, while both Western comparative studies and existing China-focused scholarship often rely on national-level typologies, such macro-level analyses tend to obscure subnational heterogeneity. The characterization of vocational schools as mere labor agencies—though grounded in legitimate concerns about student intern exploitation and uneven training quality—fails to capture the full complexity of the system. Given the decentralized governance structure described earlier, considerable variation across schools is not only plausible but empirically evident. Drawing on a qualitative study of 25 vocational schools in Guangdong Province, I identify four distinct skill formation models, only one of which aligns closely with the labor agency archetype. This diversity challenges monolithic portrayals of the system and calls for more nuanced, context-sensitive analyses.

Third, and perhaps most critically, my findings reveal that despite the state's withdrawal from operational oversight, employers have actively filled the void, becoming central actors in shaping training content and delivery through informal school-firm partnerships. This stands in stark contrast to the commonly held view that China's vocational education follows a predominantly school-based model with minimal employer engagement. Although no formal legal framework compels firms to participate in pre-employment training, de facto apprenticeship arrangements have emerged organically across many regions—well before the current state-led apprenticeship reforms were introduced. These informal collaborations reflect a broader pattern in China's reform trajectory, where decentralized experimentation drives institutional innovation (Friedman & Kuruvilla, 2015). However, when it comes to the provision of public goods such as skills, the absence of industry-wide coordination mechanisms may ultimately constrain the scalability and sustainability of these bottom-up initiatives, posing a structural challenge to systemic reform.

In sum, the Chinese VET system cannot be accurately understood through the lens of centralized state control or rigid typologies derived from national-level abstractions. Instead, it is better conceptualized as a fragmented, adaptive, and employer-influenced field shaped by local agency, institutional informality, and uneven development—a reality that demands revised theoretical frameworks and more granular empirical investigation.

Appendix A: Estimated Proportion of Vocational Graduates in the Total Urban Workforce (2000–2023)

Year	*Secondary vocational graduates (thousand)*	*Tertiary vocational graduates (thousand)*	*Total graduates from vocational schools (thousand)*	*Estimated employment rate (%)*	*Employed graduates (thousand)*	*Total urban workforce (thousand)*	*Proportion of employed vocational education graduates in the total urban workforce (%)*
2000	4767	179	4946	80	3957	720850	0.55
2001	4306	193	4499	80	3599	727970	0.49
2002	3801	277	4078	80	3262	732800	0.45
2003	3464	948	4412	80	3530	737360	0.48
2004	3592	1195	4787	80	3830	742640	0.52
2005	4182	1602	5784	80	4627	746470	0.62
2006	4791	2048	6839	80	5471	749780	0.73
2007	5309	2482	7791	80	6233	753210	0.83
2008	5807	2863	8670	80	6936	755640	0.92
2009	6249	2856	9105	80	7284	758280	0.96
2010	6650	3164	9814	80	7851	761050	1.03
2011	6600	3285	9885	80	7908	764200	1.03
2012	6746	3209	9955	80	7964	767040	1.04
2013	6744	3187	9931	80	7945	769770	1.03
2014	6229	3180	9409	80	7527	772530	0.97
2015	5679	3223	8902	80	7122	774510	0.92
2016	5336	3298	8634	80	6907	776030	0.89
2017	4969	3516	8485	80	6788	776400	0.87
2018	4873	3665	8538	80	6830	775860	0.88
2019	4935	3638	8573	80	6858	774710	0.89

H. Zhang, *Upskilling the Chinese workforce?*,
https://doi.org/10.1007/978-981-95-2656-7

Year	*Secondary vocational graduates (thousand)*	*Tertiary vocational graduates (thousand)*	*Total graduates from vocational schools (thousand)*	*Estimated employment rate (%)*	*Employed graduates (thousand)*	*Total urban workforce (thousand)*	*Proportion of employed vocational education graduates in the total urban workforce (%)*
2020	4849	3767	8616	80	6892	750640	0.92
2021	3754	3984	7738	80	6190	746520	0.83
2022	3993	4948	8940	80	7152	733510	0.98
2023	4155	5533	9687	80	7750	740410	1.05

Data Sources: Data on the total urban workforce, as well as the number of secondary and tertiary vocational graduates, were obtained from the National Bureau of Statistics of China (NBSC) annual statistical bulletins throughout the relevant years. Employment rates were estimated as 80% based on published reports from the Ministry of Education and the Ministry of Human Resources and Social Security, as well as the author's own research in different schools

Acknowledgments Some material in this book has previously appeared in a peer-reviewed journal. In particular, Chapter 5 and part of Chapter 1 is based on the article: Zhang, H. (2021). An Institutional Dilemma in China's Skills-development System: Evidence from Two Apprenticeship Reforms. *The China Quarterly*, *248*(1), 1103–1128. https://doi.org/10.1017/S030574102100059X. Reprinted with permission from Cambridge University Press.

Appendix B: Interview List

#	*Organization*	*Title*	*Time*
I: 1	Ministry of Education (MOE)	Deputy department head	November-15
I: 2	Ministry of Human Resources and Social Security (MOHRSS)	Deputy department head	February-16
I: 3	MOE and MOHRSS	Former official	February-16
I: 4	MOHRSS—Occupational Skill Testing Authority	Former director	May-16
I: 5	Guangdong Department of Education (GDDOE)	Official	November-15
I: 6	GDDOE	Department head	March-16
I: 7	GDDOE	Department head	March-16
I: 8	Guangdong Department of Human Resources and Social Security (GDDOHRSS)	Bureau head	December-15
I: 9	Shunde Bureau of Education (SDDOE)	Deputy head	January-15
I: 10	SDDOE—Office of School Affairs	Official	January-15
I: 11	Dongguan Bureau of Administration	Office head	May-16
I: 12	SDDOE—Special Committee for the Promotion of Vocational Education	Official	January-15

H. Zhang, *Upskilling the Chinese workforce?*,
https://doi.org/10.1007/978-981-95-2656-7

#	*Organization*	*Title*	*Time*
I: 13	Guangdong Academy of Education (GDAE)—vocational education institute	Director	January-16
I: 14-16	GDAE—vocational education Institute	Deputy director Official Director	January-16
I: 17	GDAE—vocational education institute	Deputy director	January-16
I: 18	GDAE—vocational education institute	Official	January-16
I: 19	GDAE	Deputy head	March-16
I: 20	GDAE—vocational education institute	Director	March-16
I: 21-25	GDAE—vocational education institute	Consultant Official Consultant Consultant Consultant	March-16
I: 26	GDAE—vocational education institute	Official	March-16
I: 27	GDDOE—Special Committee for Modern Apprenticeship	Director	May-16
I: 28	Guangzhou Bureau of Human Resources and Social Security—Occupational Skill Testing Authority	Former director	April-16
II: 1	Guangzhou Institute of Technology (GZIT)	President	December-14
II: 2	GZIT	Official	December-14
II: 3	GZIT—Employment Promotion Center	Director	January-15
II: 4-6	Guangdong Industry Polytechnic (GDIP)—Personnel Department	School president Head Deputy head	November-13
II: 7	GDIP	Deputy president	January-15
II: 8	GDIP—Department of Mechatronics	Chair	April-15
II: 9	GDIP	Deputy president	April-15
II: 10	GDIP—Office of International Affairs	Director	April-15
II: 11	Shunde Liangqiuju Vocational and Technical School (SDLVTS)	President	January-15

#	*Organization*	*Title*	*Time*
II: 12	Guangdong Construction Polytechnic (GDCP)	President	December-15
II: 13	GDCP	Provost	December-15
II: 14	GDCP	Vice president	March-16
II: 15-21	GDCP	Vice president President Visitor GDDOE official GDAE official GDAE official Teacher	March-16
II: 22-44	GDCP	Teachers and students	March-16
II: 45-46	GDCP	Teacher President	March-16
II: 47	GDCP	President	May-16
II: 48	Qingyuan Polytechnic (QYP)—Department of Beauty	Director	December-15
II: 49	QYP	Provost	December-15
II: 50	QYP	Official	December-15
II: 51	QYP	Provost	December-15
II: 52	QYP—Department of Beauty	Teacher	December-15
II: 53-55	QYP—Department of Beauty	Teacher Teacher Chair	December-15
II: 56-57	QYP—Department of Beauty	Chair Former school president	March-16
II: 58-64	Guangzhou Health Science College (GZHSC)	Association official Peer school official Program director Peer school official Firm representative Association official Firm representative	December-15
II: 65	GZHSC—Department of Medical Technology	Beauty program director	January-16
II: 66	GZHSC—Department of Medical Technology	Deputy director	January-16
II: 67-74	GZHSC	Association official Program director Peer school official Firm representative Assistant president Peer school official Natural Beauty manager Department chair	January-16

#	*Organization*	*Title*	*Time*
II: 75	GZHSC—Department of Medical Technology	Beauty program director	May-16
II: 76	GZHSC—Department of Medical Technology	Beauty program director	May-16
II: 77	GZHSC	Vice president	May-16
II: 78	GZHSC	Graduate	May-16
II: 79	Guangzhou Railway Polytechnic (GZRP)—Department of Information Engineering	Director	December-15
II: 80	GZRP—Department of Information Engineering	Deputy director	December-15
II: 81-84	GZRP	Students	December-15
II: 85	GZRP	Teacher and class advisor	December-15
II: 86-87	GZRP	Students	December-15
II: 88	Guangdong Communication Polytechnic (GDCP)—School of Automobile Application	Director	March-16
II: 89-90	GDCP—School of Rail Transit	Students	March-16
II: 91-92	GDCP—School of Transportation Management	Students	March-16
II: 93	Guangzhou Panyu Polytechnic (GZPYP)	Party secretary	March-16
II: 94	GZPYP	President	March-16
II: 95-97	GZPYP—Official of International Affairs	Director Official Teacher	March-16
II: 98-101	GZPYP	Students	March-16
II: 102	GZPYP	Official	March-16
II: 103	Guangdong Mechanical and Electrical College (GDMEC)—School of Automobile	Deputy dean	March-16
II: 104	GDMEC—School of Automobile	Deputy dean	March-16
II: 105	GDMEC—School of Automobile	Teacher	March-16
II: 106	GDMEC—School of Automobile	Teacher	March-16
II: 107-114	GDMEC—School of Automobile	Students	March-16

#	*Organization*	*Title*	*Time*
II: 115	Guangdong Polytechnic of Science and Technology (GDPST)—Guangzhou College	Director	March-16
II: 116-117	GDPST—School of Mechatronics	Assistant president President	March-16
II: 118	GDPST	Vice provost	
II: 119	GDPST	Vice president	
II: 120-123	GDPST	Students	March-16
II: 124-129	GDPST	President Vice provost Vice president Teachers	March-16
II: 130	GDPST	Student	April-16
II: 131-137	GDPST	Students	April-16
II: 138	Guangdong Vocational College of Post and Telecom (GDVCPT)	Vice provost	March-16
II: 139	GDVCPT—Department of Economics and Business	Chair	March-16
II: 140	GDVCPT	President	March-16
II: 141-143	GDVCPT	Vice president Assistant president Official	March-16
II: 144-148	GDVCPT	Vice president Department chair Program director Teachers	March-16
II. 149-152	GDVCPT	Consultant Consultant State official State official	March-16
II: 153	Shunde Vocational and Technical School (SDVTS)—Department of Automobile	Teacher	April-16
II: 154	Guangzhou Vocational School of Light Industry (GZVSLI)	President	April-16
II: 155	GZVSLI—Department of Auto Manufacturing	Chair	April-16
II: 156-59	GZVSLI	President Department chair Officials	April-16

#	*Organization*	*Title*	*Time*
II: 160	Shenzhen Polytechnic (SZP)—School of Applied Chemistry and Biotechnology	Vice dean	May-16
II: 161	SZP—School of Applied Chemistry and Biotechnology	Vice dean	May-16
II: 162	SZP—School of Management	Student	May-16
II: 163	Dongguan Electromechanics Engineering School (DGEES)	Vice president	May-16
II: 164	DGEES	Provost	May-16
II: 165	DGEES—Institute of Robots	Teacher	May-16
II: 166	Zhuhai Health School (ZHHS)	Teacher	January-16
II: 167	ZHHS—Department of Beauty	Chair	January-16
II: 168	Zhaoqing Medical College (ZQMC)—Department of Nursing	Chair	January-16
II: 169	ZQMC—Department of Beauty	Teacher	January-16
II: 170	Zhanjiang Health School (ZJHS)—Department of Beauty	Teacher	January-16
II: 171	ZJHS—Department of Beauty	Teacher	January-16
II: 172	Guangzhou Conghua Technical School (GZCHTS)	President	January-15
II: 173	GZCHTS	Vice president	January-15
II: 174	Guangdong Lingnan #1 Technical College of Industry and Commerce (GDLNCIC)—Department of Mechatronics	Former chair	April-16
II: 175	GDLNCIC—Department of Automobile	Director	April-16
II: 176	GDLNCIC—Department of Interior Design	Student	April-16
II: 177-179	GDLNCIC—Department of Mechatronics	Students	April-16
II: 180-181	GDLNCIC—Department of Mold Manufacturing	Students	April-16
II: 182-184	GDLNCIC	Parents	April-16
II: 185	GDLNCIC	Prospective student	April-16
II: 186	Guangzhou Technician College (GZTC)	President	April-16

#	*Organization*	*Title*	*Time*
II: 187	GZTC—Center of Occupational Skill Training and Testing	Director	April-16
II: 188	GZTC—Department of Automobile	Chair	April-16
II: 189	Guangdong Machinery Technician College (GDMTC)	Vice president	April-16
II: 190	GDMTC	WorldSkills Competition Champion	April-16
II: 191	Baiyun Technician College of Business and Technology (BYTCBT)—vocational education Institute	Director	April-16
II: 192	BYTCBT—Department of Tourism	Vice chair	April-16
II: 193	BYTCBT—Department of Mechatronics	Chair	May-16
II: 194	BYTCBT—Department of Mechatronics	Teacher	May-16
II: 195	BYTCBT—Department of Mechatronics	Program director	May-16
II: 196	BYTCBT—Department of Mechatronics	Career advisor	May-16
II: 197	BYTCBT—Department of Architecture	Student	May-16
II: 198	BYTCBT	Dormitory houseparent	May-16
II: 199	Beijing Union Cadre School (BJUCS)	Vice president	November-15
II: 200	BJUCS	Teacher	November-15
III: 1	Gree Electric Appliances, Inc.	Recruiting manager	January-15
III: 2-3	Hong Kong Yagelo Cosmetics Group	President General manager	December-15
III: 4	Horizontal Information Technology, Inc.	Chief engineer	January-16
III: 5	Natural Beauty Group	Firm teacher	January-16
III: 6-8	Hong Kong Yagelo Cosmetics Group	President General manager Training institute director	January-16
III: 9	Ellehuis, Co., Ltd.—Training Center	Director	January-16
III: 10-12	Ellehuis, Co., Ltd.	Class advisor Salon manager Worker	January-16

#	*Organization*	*Title*	*Time*
III: 13	Ellehuis, Co., Ltd.	Salon manager	January-16
III: 14	Ellehuis, Co., Ltd.	Class advisor	January-16
III: 15	Ellehuis, Co., Ltd.	President	March-16
III: 16-18	Ellehuis, Co., Ltd.	President Vice president CHO	March-16
III: 19	Ellehuis, Co., Ltd.	Salon manager	March-16
III: 20	Ellehuis, Co., Ltd.	Worker	March-16
III: 21	Ellehuis, Co., Ltd.—Training Center	Deputy director	March-16
III: 22	Ellehuis, Co., Ltd.	Worker	March-16
III: 23-24	Ellehuis, Co., Ltd.	Workers	March-16
III: 25-26	Ellehuis, Co., Ltd.	CHO Collaborative school president	March-16
III: 27	Guangzhou Otis, Inc.	Former HR director	March-16
III: 28	Tiandi Huayu Logistics Group	HR director	March-16
III: 29	Tiandi Huayu Logistics Group	HR officer	March-16
III: 30	Zhongtian Group #7 Construction Corporation	General engineer	March-16
III: 31	Zhongtian Group #7 Construction Corporation	Manager	March-16
III: 32-38	Zhongtian Group #7 Construction Corporation	Workers	March-16
III: 39	Guangzhou Seagull Kitchen and Bath Products Co., Ltd.	HR manager	March-16
III: 40	Guangzhou Seagull Kitchen and Bath Products Co., Ltd.	Training manager	March-16
III: 41	Guangzhou Seagull Kitchen and Bath Products Co., Ltd.	Trainer	March-16
III: 42-43	Guangzhou Seagull Kitchen and Bath Products Co., Ltd.	HR manager Training manager	March-16
III: 44	Zhuhai Europe-Asia Auto Tech Co., Ltd.	Vice president	March-16
III: 45	Zhuhai Europe-Asia Auto Tech Co., Ltd.	President	March-16
III: 46-47	Zhuhai Europe-Asia Auto Tech Co., Ltd.	President Vice president	March-16

#	*Organization*	*Title*	*Time*
III: 48-53	Zhuhai Europe-Asia Auto Tech Co., Ltd.	Industrial association officials Industrial association official President Vice president	April-16
III: 54	Guangdong STS Telecom Tech, Inc.	Manager	March-16
III: 55	Guangzhou Telecom 10000, Inc.	Training manager	March-16
III: 56-61	Guangzhou Telecom 10000, Inc.	Manager Training manager School official Workers	March-16
III: 62-64	Guangzhou Telecom 10000, Inc.	Training manager School president Manager	March-16
III: 65	Guangzhou Telecom 10000, Inc.	Training manager	April-16
III: 66	Guangxin Communications Services Co., Ltd.	Manager	March-16
III: 67	Jabil Electronics (Guangzhou), Inc.	HR manager	March-16
III: 68	FAW-VW Automotive (Foshan) Co., Ltd	Workshop head and union committee member	April-16
III: 69	FAW-VW Automotive (Foshan) Co., Ltd	Workshop head and union committee member	April-16
III: 70-72	Lenuan Heating Appliances Co., Ltd.	Vice president Factory head Assistant president	April-16
III: 73	Guangzhou Radio Group	Union chair	April-16
III: 74	GAC-Toyota Motor Co., Ltd.	HR official	April-16
III: 75	Guangzhou Jiehe Electric Tech, Inc.	General manager	May-16
III: 76	Guangzhou Jiehe Electric Tech, Inc.	Sales manager	May-16
III: 77	Guangzhou Jiehe Electric Tech, Inc.	Technician	May-16
III: 78-79	Dongguan Janus Co., Ltd	Training center director Training center manager	May-16 May-16
IV: 1	Internet of Things Association of Guangdong	Training center director	December-15
IV: 2	Internet of Things Association of Guangdong	Training center director	March-16

#	*Organization*	*Title*	*Time*
IV: 3	Chinese Academy of Labor and Social Security—Vocational Training Department	Research associate	December-15
IV: 4	Chinese Academy of Labor and Social Security—Vocational Training Department	Director	February-16
IV: 5	Beijing Normal University—Institute of Vocational and Adult Education	Professor	December-15
IV: 6	Beijing Normal University—Institute of Vocational and Adult Education	Professor	May-16
IV: 7	Guangdong Health Vocational Education Association	President	December-15
IV: 8	Shunde Association for the Promotion of Vocational Education	Deputy secretary general	January-16
IV: 9	Huizhou Beauty and Cosmetic Association	President	January-16
IV: 10	South China University of Technology	Professor	March-16
IV: 11	Guangzhou Federation of Trade Unions	Former chair	March-16
IV: 12	British Council (Beijing)	Educational project manager	March-16
IV: 13	British Council (Guangzhou)	Educational project manager	April-16
IV: 14	Shunde E-Commerce Association	Department manager	April-16
IV: 15	Guangdong Enterprise Confederation—Employer Service Department	Deputy director	April-16

References

Althusser, L. (2006). Ideology and ideological state apparatuses. In A. Sharma & A. Gupta (Eds.), *The anthropology of the state: A reader* (pp. 86–111). Blackwell.

Ashton, D., et al. (1999). *Economies, education and training for development in East Asia: The political economy of skill formation in newly industrialised economies.* Routledge.

Ashton, D., Sung, J., & Turbin, J. (2000). Towards a framework for the comparative analysis of national systems of skill formation. *International Journal of Training and Development, 4*(1), 8–25.

Ashton, D., Sung, J., & Turbin, J. (2010). The evolution of education and training strategies in Singapore, Taiwan and S. Korea: A development model of skill formation. *Journal of Education and Work, 15*(1), 5–30.

Barney, J. (1991). Firm resources and sustained competitive advantage. *Journal of Management, 17*(1), 99–120.

Becker, G. S. (1993). *Human capital: A theoretical and empirical analysis, with special reference to education* (3rd ed.). The University of Chicago Press.

Boyd, B. K., Bergh, D. D., & Ketchen, D. J. (2010). Reconsidering the reputation-performance relationship: A resource-based view. *Journal of Management, 36*(3), 588–609.

Boyer, R. (1990). *The regulation school: A critical introduction.* Columbia University Press.

Cappelli, P. (1995). Rethinking employment. *British Journal of Industrial Relations, 33*(4), 563–602.

Chan, K. W. (2010). A China paradox: Migrant labor shortage amidst rural labor supply abundance. *Eurasian Geography and Economics, 51*(4), 513–530.

H. Zhang, *Upskilling the Chinese workforce?*,
https://doi.org/10.1007/978-981-95-2656-7

Chan, J., Pun, N., & Selden, M. (2015). Interns or workers? China's student labor regime. *Asian Studies, 1*(1), 69–98.

Chen, T. (2015). Gaodeng jiaoyu xueke zhuanye mulu: wenti yu luoji (Catalogue of Disciplines and Specialties in Higher Education: Problems and Logic). *Journal of Southwest Jiaotong University, 16*(3), 43–49.

Child, J. (1994). *Management in China during the age of reform.* Cambridge University Press.

Cooke, F. L. (2005). Vocational and enterprise training in China: Policy, practice and prospect. *Journal of the Asia Pacific Economy, 10*(1), 26–55.

Corley, K., & Gioia, D. (2000). The rankings game: Managing business school reputation. *Corporate Reputation Review, 3*, 319–333.

Crouch, C., Finegold, D., & Sako, M. (1999). *Are skills the answer? The political economy of skill creation in advanced industrial societies.* Oxford University Press.

Culpepper, P. D. (1999a). Still a model for the industrialization countries? In P. D. Culpepper & D. Finegold (Eds.), *The German skills machine: Sustaining comparative advantage in a global economy* (pp. 1–36). Berghahn Books.

Culpepper, P. D. (1999b). The future of the high-skill equilibrium in Germany. *Oxford Review of Economic Policy, 15*(1), 43–59.

Davis-Blake, A., & Uzzi, B. (1993). Determinants of employment externalization: A study of temporary workers and independent contractors. *Administrative Science Quarterly, 38*(2), 195–223.

Delebarre, J. (2016). Apprenticeships policy, England 2015. *House of Commons Library Briefing Paper.*

Doeringer, P. B., & Piore, M. J. (1971). *Internal labor markets and manpower analysis.* Heath Lexington Books.

Dougherty, K. J., & Bakia, A. F. (1999). *The new economic development role of the community college.* Community College Research Center.

Durden, G. R., & Yang, G. (2006). Higher vocational education in China: A preliminary critical review of developments and issues in Liaoning Province. *Journal of European Industrial Training, 30*(8), 622–638.

Farrell, D., & Grant, A. J. (2005). China's looming talent shortage. *McKinsey Quarterly, 4*, 70–79.

Feldman, M., Francis, J., & Bercovitz, J. (2005). Creating a cluster while building a firm: entrepreneurs and the formation of industrial clusters. *Regional Studies, 39*(1), 129–141.

Feng, X. (2019). Regulating labour dispatch in China: A cat-and-mouse game. *China Information, 33*(1), 88–109.

Finegold, D., & Soskice, D. (1988). The failure of training in Britain: Analysis and prescription. *Oxford Review of Economic Policy, 4*(3), 21–53.

Finegold, D., & Wagner, K. (2002). Are apprenticeships still relevant in the 21st century? A case study of changing youth training arrangements in German Banks. *Industrial and Labor Relations Review, 55*(4), 667–685.

Friedman, E. (2014a). *Insurgency trap: Labor politics in postsocialist China*. Cornell University Press.
Friedman, E. (2014b). Economic development and sectoral unions in China. *Industrial and Labor Relations Review, 67*(2), 481–503.
Friedman, E., & Kuruvilla, S. (2015). Experimentation and decentralization in China's labor relations. *Human Relations, 68*(2), 181–195.
Friedman, E., & Lee, C. K. (2010). Remaking the world of Chinese labour: A 30-year retrospective. *British Journal of Industrial Relations, 48*(3), 507–533.
Furth, D. (1985). *Education and training after basic schooling*. OECD.
Gallagher, M. E., & Dong, B. (2011). Legislating harmony: Labor law reform in contemporary China. In S. Kuruvilla, C. K. Lee, & M. E. Gallagher (Eds.), *From iron rice bowl to informalization*. Cornell University Press.
Gallagher, M. E., Lee, C. K., & Kuruvilla, S. (2011). Introduction and argument. In S. Kuruvilla, C. K. Lee, & M. E. Gallagher (Eds.), *From iron rice bowl to informalization* (pp. 1–14). Cornell University Press.
Green, A. (1991). The reform of post-16 education and training and the lessons from Europe. *Journal of Education Policy, 6*(3), 327–339.
Green, F. (1992). On the political economy of skill in the advanced industrial nations. *Review of Political Economy, 4*, 413–435.
Hall, P., & Soskice, D. (2001). An introduction to varieties of capitalism. In P. Hall & D. Soskice (Eds.), *Varieties of capitalism: The institutional foundations of comparative advantage*. Oxford University Press.
Hashimoto, M. (1979). Bonus payments, on-the-job training, and lifetime employment in Japan. *Journal of Political Economy, 87*(5), 1086–1104.
Hashimoto, M. (1981). Firm-specific human capital as a shared investment. *American Economic Review, 71*(3), 475–482.
Hashimoto, M., & Yu, B. T. (1980). Specific capital, employment contracts, and wage rigidity. *The Bell Journal of Economics, 11*(2), 536–549.
Helmke, G., & Levitsky, S. (2004). Informal institutions and comparative politics: A research agenda. *Perspectives on Politics, 2*(4), 725–740.
Howell, J., & Pringle, T. (2019). Shades of authoritarianism and state–labour relations in China. *British Journal of Industrial Relations, 57*(2), 223–246.
Hurst, W. (2009). *The Chinese worker after socialism*. Cambridge University Press.
Klein, B., Crawford, R., & Alchian, A. (1978). Vertical integration, appropriable rents and the competitive contracting process. *Journal of Law and Economics, 21*(2), 297–326.
Knoke, D., & Ishio, Y. (1994). Occupational training, unions, and internal labor market. *American Behavioral Scientist, 37*(7), 992–1016.
Kochan, T. A., Katz, H. C., & McKersie, R. B. (1986). *The transformation of American industrial relations*. Basic Books.

Kong, W. (2015). On the problems of Chinese professional qualification certificates system and countermeasures (Woguo Zhiye Zige Zhengshu Zhidu De Wenti Yu Duice). *Journal of Henan Institute of Science and Technology, 8*, 10–14.

Kuruvilla, S. (2018). From cautious optimism to renewed pessimism: Labor voice and labor scholarship in China. *Industrial and Labor Relations Review, 71*(5), 1013–1028.

Kuruvilla, S., & Chua, R. (2000). How do nations increase workforce skills? Factors influencing the success of the Singapore skills development system. *Global Business Review, 1*(1), 11–47.

Kuruvilla, S., & Zhang, H. (2016). Labor unrest and incipient collective bargaining in China. *Management and Organization Review, 12*(1), 159–187.

Kuruvilla, S., Erickson, C. L., & Hwang, A. (2002). An assessment of the Singapore skills development system: Does it constitute a viable model for other developing countries? *World Development, 30*(8), 1461–1476.

Kwan, F. (2009). Agricultural labour and the incidence of surplus labour: Experience from China during reform. *Journal of Chinese Economic and Business Studies, 7*(3), 341–361.

Lai, M., & Lo, L. N. K. (2006). Decentralization and social partnership: The development of vocational education at Shanghai and Shenzhen in China. *Educational Research for Policy and Practice, 5*(2), 101–120.

Leborgne, D., & Lipietz, A. (1988). New technologies, new modes of regulation: Some spatial implications. *Society and Space, 6*, 263–280.

Lepak, D. P., & Snell, S. A. (1999). The human resource architecture: Toward a theory of human capital allocation and development. *Academy of Management Review, 24*(1), 31–48.

Li, Y., & Sheldon, P. (2010). HRM lives inside and outside the firm: Employers, skill shortages and the local labour market in China. *The International Journal of Human Resource Management, 21*(12), 2173–2193.

Li, Y., & Sheldon, P. (2014). Collaborations between Foreign-invested Enterprises and China's Vocational Education Schools: Making the system work amid localised skill shortages. *Journal of Vocational Education & Training, 66*(3), 311–329.

Lieberman, M. B., & Montgomery, D. B. (1988). First-mover advantages. *Strategic Management Journal, 9*(S1), 41–58.

Liu, X. (2015). How institutional and organizational characteristics explain the growth of contingent work in China. *Industrial and Labor Relations Review, 68*(2), 372–397.

Locke, R. (1992). The demise of the national union in Italy: Lessons for industrial relations theory. *Industrial and Labor Relations Review, 45*(2), 229–249.

Locke, R., Kochan, T., & Piore, M. (1995). Reconceptualizing Comparative Industrial Relations: Lessons from International Research. *International Labour Review, 134*(2), 139–161.

Lucas, R. E. (1988). On the Mechanics of Economic Development. *Journal of Monetary Economics, 22*, 3–42.
Luo, B. N., et al. (2022). The human resource architecture model: A twenty-year review and future research directions. In *Strategic Human Resource Management and Organizational Effectiveness* (pp. 17–54).
Lüthje, B., Luo, S., & Zhang, H. (2013). *Beyond the Iron Rice Bowl: Regimes of Production and Industrial Relations in China*. Campus Verlag.
Marx, K. (1978). Preface to "A contribution to the critique of political economy". In R. Tucker (Ed.), *The Marx-Engels reader* (pp. 4–5). Norton. (Original work published 1859).
Monks, J., & Ehrenberg, R. G. (1999). U.S. News & World Report's college rankings: Why do they matter? *Change, 31*(6), 42–51.
Müller, A. (2024). Cooperation between colleges and companies: Vocational education, skill mismatches and China's turnover problem. *The China Quarterly, 260*, 986–1004.
Murphy, R., & Johnson, D. (2009). Education and development in China: Institutions, curriculum and society. *International Journal of Educational Development, 29*(5), 447–453.
Nee, V., & Ingram, P. (1998). Embeddedness and beyond: Institutions, exchange and social structure. In M. Brinton & V. Nee (Eds.), *New institutionalism in sociology* (pp. 19–45). Russell Sage Foundation.
Nee, V., & Opper, S. (2012). *Capitalism from below: Markets and institutional change in China*. Harvard University Press.
Ngok, K., & Guo, W. (2008). The quest for world class universities in China: Critical reflections. *Policy Futures in Education, 6*(5), 545–557.
Olney, S. L. (1994). *Unions in a changing world: Problems and prospects in selected industrialized countries*. International Labour Office.
Olson, M. (1965). *The logic of collective action: Public goods and the theory of groups*. Harvard University Press.
Osterman, P. (1982). Employment structures within firms. *British Journal of Industrial Relations, 20*(3), 349–361.
Osterman, P. (1987). Choice of employment systems in internal labor markets. *Industrial Relations, 26*(1), 46–67.
Osterman, P., & Batt, R. (1993). Employer-centered training for international competitiveness: Lessons from state programs. *Journal of Policy Analysis and Management, 12*(3), 456–477.
Osterman, P., & Burton, M. D. (2006). *Ports and ladders: The nature and relevance of internal labor markets in a changing world*. Unpublished manuscript.
Parker, E., & Rogers, J. (1999). Sectoral training initiatives in the US: Building blocks of a new workforce preparation system? In P. D. Culpepper & D. Finegold (Eds.), *The German skills machine: Sustaining comparative advantage in a global economy* (pp. 326–362).

Patton, M. Q. (1990). *Qualitative evaluation and research methods* (2nd ed.). Sage.
Peters, K. (2007). Business school rankings: Content and context. *Journal of Management Development, 26*(1), 49–53.
Pfeffer, J., & Fong, C. T. (2002). The end of business schools? Less success than meets the eye. *Academy of Management Learning and Education, 1*(1), 78–95.
Pfeffer, J., & Salancik, G. R. (1978). *The external control of organizations: A resource dependence perspective*. Harper and Row.
Piore, M. J. (2002). Thirty years later: Internal labor markets, flexibility and the new economy. *Journal of Management and Governance, 6*(4), 271–279.
Rosenfeld, S. A. (2000). Community college/cluster connections: Specialization and competitiveness in the United States and Europe. *Economic Development Quarterly, 14*(1), 51–62.
Schnarr, A., Yang, S., & Gleissner, K. (2008). *Vocational education and training and the labour market: A comparative analysis of China and Germany*. International Center for Technical and Vocational Education and Training (UNEVOC).
Sheldon, P., & Li, Y. (2013). Localized poaching and skills shortages of manufacturing employees among MNEs in China. *Journal of World Business, 48*, 186–195.
Smith, C., & Chan, J. (2015). Working for two bosses: Student interns as constrained labour in China. *Human Relations, 68*(2), 305–326.
Song, X. (2024). Can the two Karls – Karl Polanyi and Karl Marx meet? A critical examination of Chinese vocational education, internship regimes, and labor forms. *Critical Sociology*. Advance online publication. https://doi.org/10.1177/08969205241292751
Song, X., & Xu, D. (2024). More graduates, fewer skills? Vocational education expansion and skilled labour shortages in China. *The China Quarterly, 260*, 970–985.
Sorge, A., & Streeck, W. (1988). Industrial relations and technical change: The case for an extended perspective. In R. Hyman & W. Streeck (Eds.), *New technology and industrial relations* (pp. 14–38). Blackwell.
Stevens, M. (1999). Human capital theory and UK vocational training policy. *Oxford Review of Economic Policy, 15*(1), 16–32.
Stone, J. R., III, & Lewis, M. V. (2010). Governance of vocational education and training in the United States. *Research in Comparative and International Education, 5*(3), 274–288.
Streeck, W. (1987). *The role of the social partners in vocational training and further training in the Federal Republic of Germany*. CEDEFOP.
Streeck, W. (1989). Skills and the limits of neo-liberalism: The enterprise of the future as a place of learning. *Work, Employment & Society, 3*(1), 89–104.
Streeck, W. (1992). *Social institutions and economic performance: Studies of industrial relations in advanced capitalist economies*. Sage.

Su, Y. (2010–2011). Student workers in the Foxconn empire: The commodification of education and labor in China. *Journal of Workplace Rights, 15*(3–4), 341–362.

Thelen, K. (2004). *How institutions evolve: The political economy of skills in Germany, Britain, the United States, and Japan*. Cambridge University Press.

Tsai, L. L. (2002). Cadres, temple and lineage institutions, and governance in rural China. *The China Journal, 48*, 1–27.

Tsai, K. S. (2006). Institutions and endogenous institutional change. *World Politics, 59*, 116–141.

Tsai, L. L. (2007). Solidary groups, informal accountability, and local public goods provision in rural China. *American Political Science Review, 101*(2), 355–372.

Walder, A. G. (1988). *Communist neo-traditionalism: Work and authority in Chinese industry*. University of California Press.

Wang, X. (2014). *Jineng xingcheng de shehui jiangou [The social construction of skill formation]*. Social Sciences Academic Press.

Wang, X. (2021). *Zouxiang jineng shehui: Guojia jineng xingcheng tixi yu chanye gongren jineng xingcheng [Towards a skills society: National skill formation systems and industrial workers' skill development]*. China Workers Publishing House.

Warner, M. (1993). Human resource management "with Chinese characteristics". *International Journal of Human Resource Management, 4*(1), 45–65.

Williamson, O. E. (1981). The economics of organization: The transaction cost approach. *American Journal of Sociology, 87*(3), 548–577.

Williamson, O. E., Wachter, M. L., & Harris, J. E. (1975). Understanding the employment relation: The analysis of idiosyncratic exchange. *The Bell Journal of Economics, 6*(1), 250–278.

World Bank. (2013). *China 2030: Building a modern, harmonious, and creative society*. World Bank.

Woronov, T. E. (2012). Doing time: Mimetic labor and human capital accumulation in Chinese vocational schools. *South Atlantic Quarterly, 111*(4), 701–719.

Woronov, T. E. (2016). *Class work: Vocational schools and China's urban youth*. Stanford University Press.

Wu, X., & Ye, Y. (2018). *Technical and vocational education in China*. Springer.

Xiao, J., & Tsang, M. C. (2004). Determinants of participation and nonparticipation in job-related education and training in Shenzhen. *China. Human Resource Development Quarterly, 15*(4), 389–442.

Xu, S. (1993). Zhongguo jigong xuexiao fazhan yu gaige de lishi huigu [A historical review of the development and reform of Chinese technical schools]. *Vocational Education Research, 1994*(4), 4–7.

Xu, F. (2008). The emergence of temporary staffing agencies in China. *Comparative Labor Law and Policy Journal, 30*(3), 431–461.

Xu, J. (2023). *Reassessing vocational education in China: A perspective from PISA*. Routledge.

Xue, B. (2016). Ji qi ren chan ye jiang ying lai bao fa shi zeng zhang [The robotics industry will see explosive growth]. *Employment and Security, 4*, 14–15.

Xue, E., & Li, J. (2022). *China's vocational education reform: Explorations and analysis*. Springer.

Yang, R., & Welch, A. (2012). A world-class university in China? *The case of Tsinghua. Higher Education, 63*(5), 645–666.

Yi, H., Zhang, L., Liu, C., Chu, J., Loyalka, P., Maani, M., & Wei, J. (2013). How are secondary vocational schools in China measuring up to government benchmarks? *China & World Economy, 21*(3), 98–120.

Yi, H., Li, G., Li, L., Loyalka, P., Zhang, L., Xu, J., Kardanova, E., Shi, H., & Chu, J. (2018). Assessing the quality of upper-secondary vocational education and training: Evidence from China. *Comparative Education Review, 62*(2), 199–230.

Zhou, Y. (2013). The state of precarious work in China. *American Behavioral Scientist, 57*(3), 354–372.

Index

H. Zhang, *Upskilling the Chinese workforce?*,
https://doi.org/10.1007/978-981-95-2656

Kolibri 360 GmbH
Ferdinand-Jühlke-Straße 7
99095 Erfurt, Deutschland
produktsicherheit@kolibri360.de